American Institute of Chemical Engineers

Conference of the Process Development Division 2007

Held at the 2007 AIChE Spring National Meeting

April 22-27, 2007
Houston, Texas, USA

Printed from e-media with permission by:

Curran Associates, Inc.
57 Morehouse Lane
Red Hook, NY 12571
www.proceedings.com

ISBN: 978-1-60423-364-3

Some format issues inherent in the e-media version may also appear in this print version.

ISBN: 978-1-60423-364-3

American Institute of Chemical Engineers
Proceedings
Three Park Avenue
New York, NY 10016-5991
Phone: 212-591-8100

www.aiche.org

American Institute of Chemical Engineers

Conference of the Process Development Division
2007

TABLE OF CONTENTS

Session 13 – Tools and Best Practices in Pilot Plants I

Session 38 – Tools and Best Practices in Pilot Plants II

Session 56 – Process Intensification Topics I

Session 77 – Process Intensification Topics II

Session 93 – Heat Exchanger Fouling: A Messy Business

Session 116 – To Toll or Not to Toll: That is the Question

Session 123 – Sustainable Optimization and Troubleshooting of Chemical Processes

Emission Profiler©2005 SPRI **Calculation and Reporting Software Targets NJDEP Pilot Plant, R&D General Permit Compliance**

Elizabeth A. Dirnfeld and Frederick R. Kohler, Jr.

Schering-Plough Research Institute, 1011 Morris Avenue, Union, NJ 07083

John F. Takacs

HighPoint Software Services, Inc., P.O. Box 874, Westminster, MA 01473

INTRODUCTION

Research & Development (R&D) pharmaceutical batch processing of active pharmaceutical ingredients (API) in New Jersey provides a compliance challenge to engineers, who are often held accountable for Safety, Health & Environmental (SHE) impacts. Emission Profiler's simplicity lends itself to modeling of sequential unit operations within a smart and logical framework. Given a batch recipe and Emission Profiler, an engineer can weed out troublesome operations, apply control device removals, determine emission rate compliance, and keep a tab on HAP and other targeted pollutants. Reports can be customized by the user for both regulatory purposes and internal tracking.

Emission Profiler SYSTEM

The Emission Profiler system is a Windows-based program developed in Microsoft's Visual Basic to calculate, track and report VOC and HAP emissions in the batch production of pharmaceutical products. Emission Profiler is based on US EPA Office of Quality Planning and Standards (OAQPS) calculation methods. [1,2,3] EP maintains all individual unit operations for each experiment in the system. Control devices are also tracked in the system to calculate pre- and post-control emissions. To verify the calculations, a detailed step-by-step printout of each unit operation's calculations is available with references in the calculations to the appropriate US EPA equation. The reports in the system are used to meet state, federal and internal reporting requirements.

Figure 1. Emission Profiler Start-up Screen.

Figure 2 shows the main toolbar in Emission Profiler.

Figure 2. Emission Profiler Main Toolbar.

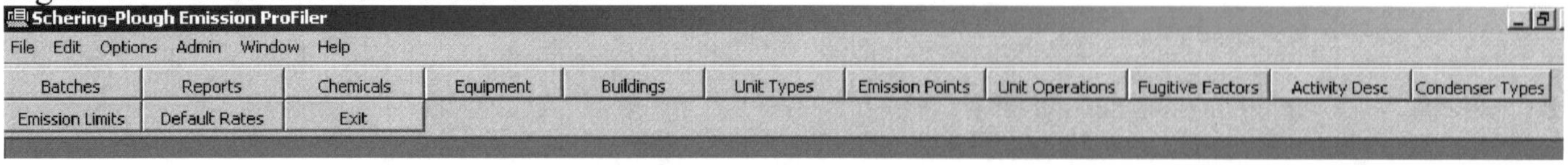

In Emission Profiler users enter batch product information, produce reports or maintain support databases. The support databases contain information that is used throughout the system. Each support databases is a separate button in the Emission Profiler toolbar. An example of the support databases for Unit Operations is shown in Figure 3. The Unit Operations table includes information on the operation type, OAQPS equation type, required input parameters and default input parameters.

Figure 3. Support Table – Unit Operations

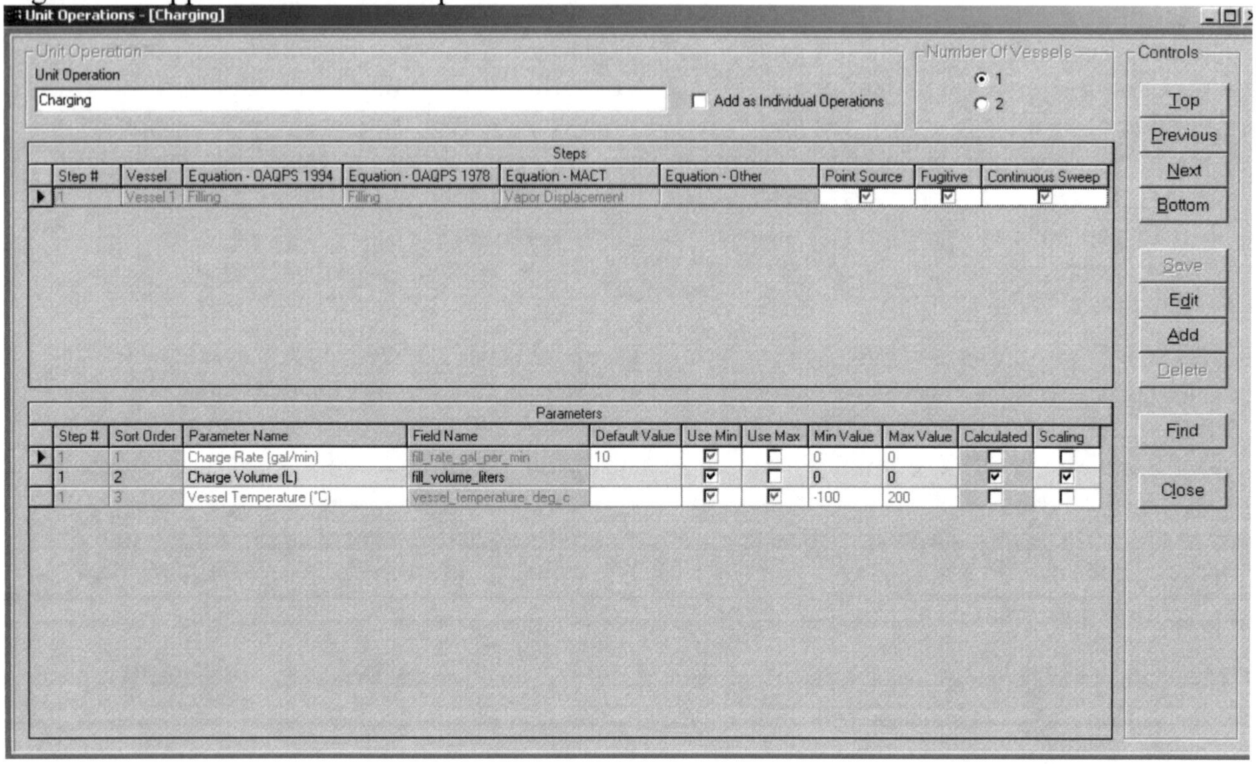

The maneuvering controls in Figure 3 are similar to those used throughout EP. The user can move through the records in the database with the Top, Previous, Next or Bottom button or use the Find button to display a list of all records in the database. To modify existing data the user selects the Edit button. The Add button allows the user to enter a new unit operation type. After editing or adding new information, the user presses the Save button to save all of the new information. The Delete button allows the user to remove records.

The entry of batch experiment information is in the Batches module. Pressing the Batches button displays the Load Experiment screen shown in Figure 4. This screen allows the user to select a year to work with. When a year is selected (highlighted), all of the existing experiments in the selected year are displayed in the bottom grid. In the experiments grid, the user can select an experiment to work with or add/remove experiments.

Figure 4. Batch Records Year / Experiment Selection Screen.

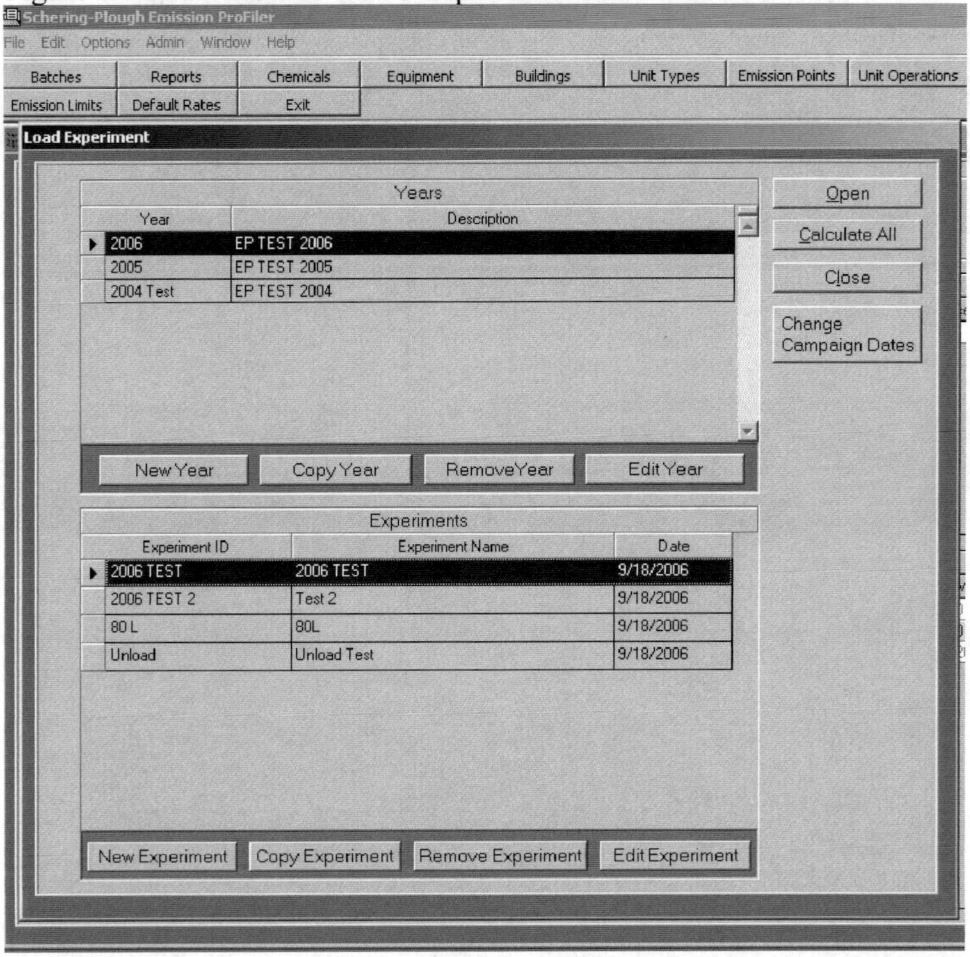

After selecting an experiment to work with, the user presses the Open button. Opening an experiment (Figure 5) displays the campaigns screen. The experiment information consists of a flip card that supplies detailed information on the campaign, individual batch unit operations, chemicals used in the process, control devices and emissions information at the batch record level. Figure 5 shows the campaigns flip card. On the campaigns flip card the experiment start and stop time is recorded along with default information for any process condensers present and a data field for adding specific comments about the experiment campaign.

Figure 5. Campaigns Screen for a Selected Experiment.

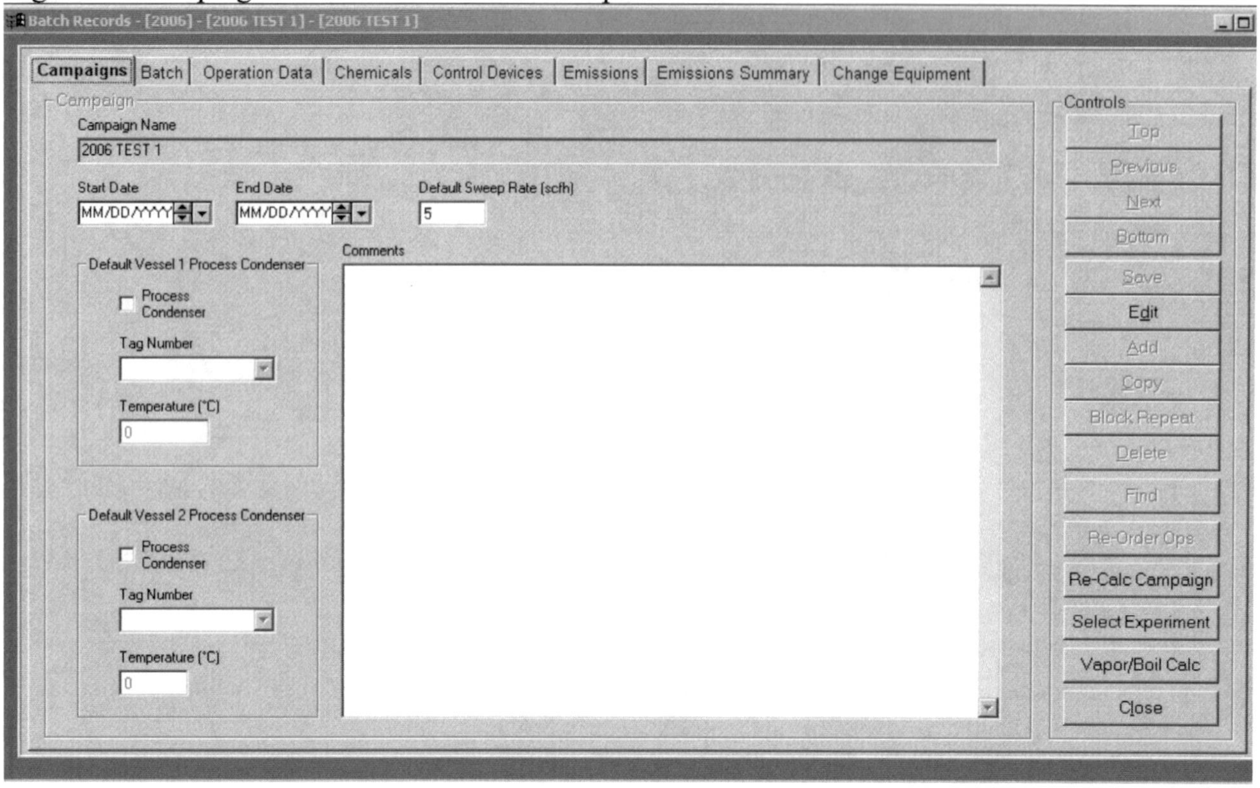

The Copy button at the campaign level allows the user to a make a complete copy of the selected campaign including all batch record steps, chemicals and control devices.

Selecting the Batch flip card (Figure 6) displays a list of the individual unit operations that make up the entire batch record cycle. To add a new step to this batch record, the user presses the Add button. The Copy button will copy an existing batch record with all of its information and add it as a new record at the end of the batch record cycle. There is also a block repeat function that allows selecting one or more unit operations and copying them as many times as necessary. If the order of the unit operations in the batch record needs to be changed, the operations can be rearranged using the Re-Order Ops button.

To look at the detailed information about an individual unit operation in the batch record, the user clicks on a record in the grid. The selected record then becomes highlighted. Once a record is highlighted, the user can look at detailed input and emissions information for that unit operation on the following flip cards.

Figure 6. Batch Record Grid

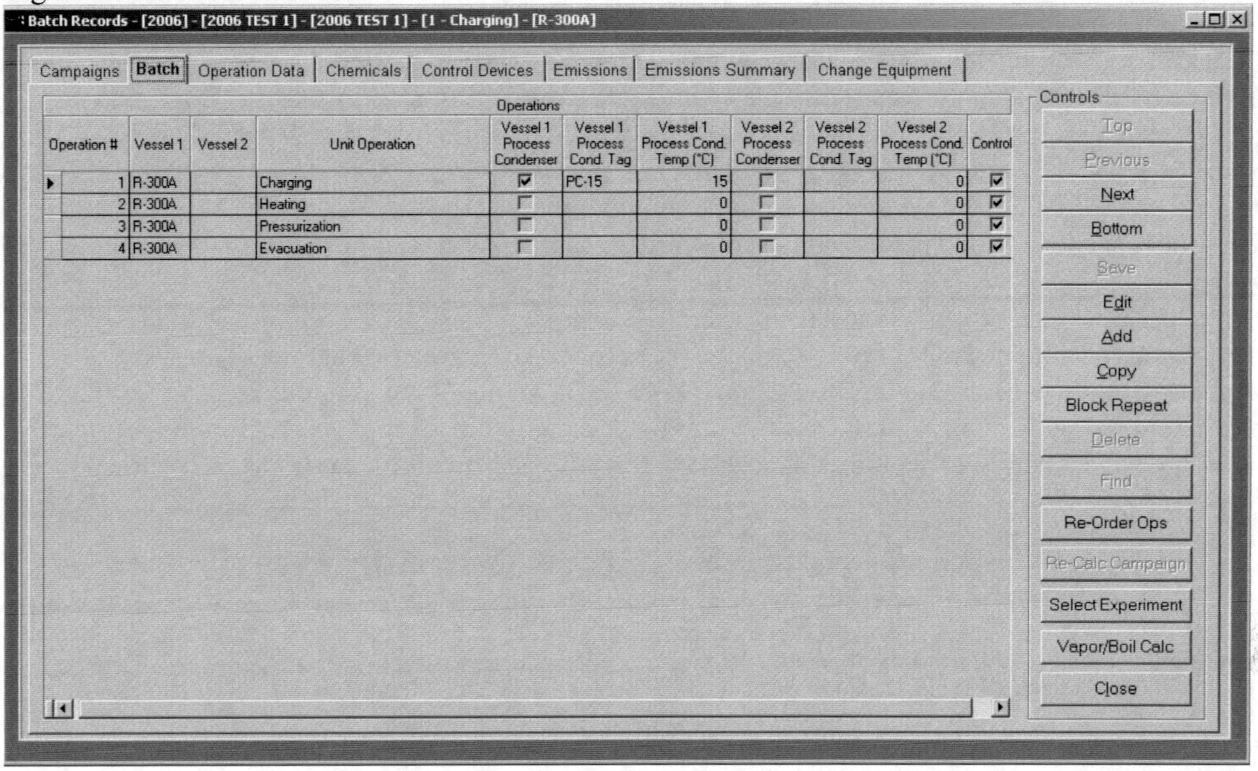

Selecting the Operation Data flip card (Figure 7) displays the input data for the individual unit operation.

Figure 7. Unit Operation Op Data Input Screen.

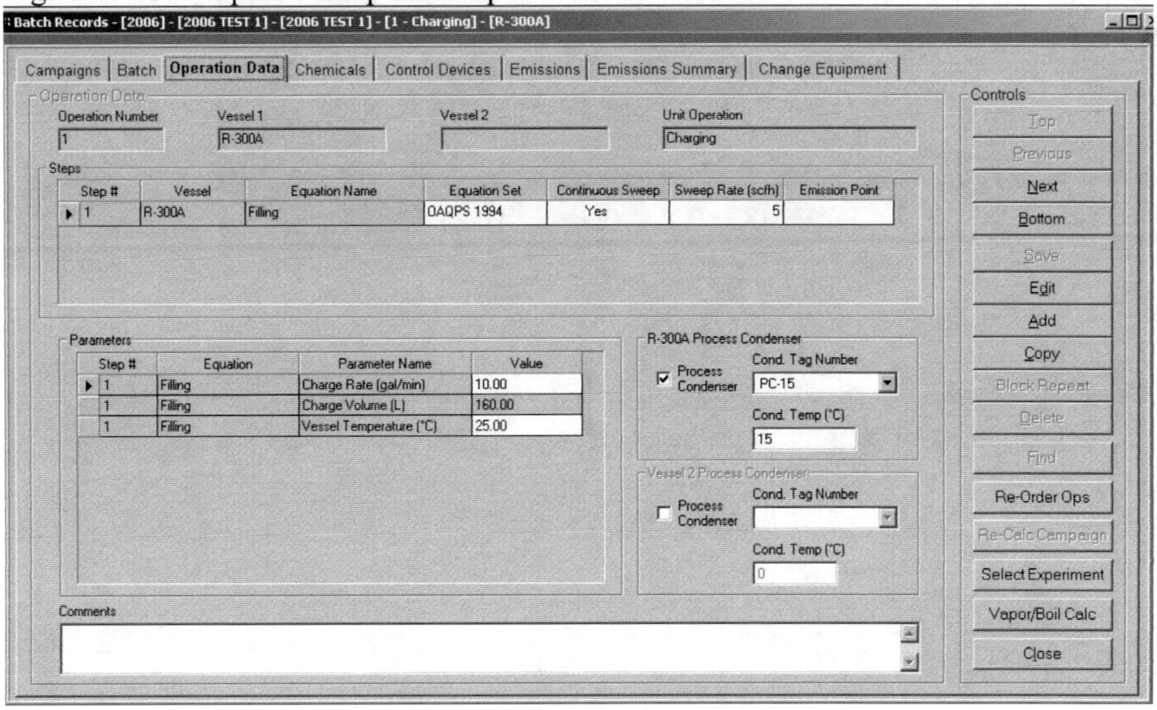

The unit operation input data fields will vary depending on the required information for a unit operation. Figure 8 displays an example for a vessel Charging operation. The required input fields are vessel charge rate and the exit temperature from the vessel. The amount of material charged to the vessel is determined on the Chemicals flip card. On the Operation Data screen the user is also able to indicate if a process condenser is present and if a continuous gas sweep takes place during this unit operation. The Steps grid at the top part of the screen, shows unit operations that composed this step. Some unit operations can have more than one step. In the case of a Charging operation, there is only one step. In this grid the OAQPS equation set used is documented along with information about a continuous gas sweep on the operation step.

Selecting the Chemicals flip card (Figure 8) displays the screen that shows the chemicals used in the selected unit operation. The top grid on this screen shows the contents of the vessel that is the results of any previous unit operation along with the activity in this unit operation. The Activity Summary grid at the bottom of the page, shows the chemical transactions for this specific unit operation.

Figure 8. Unit Operation Chemical Usage Screen.

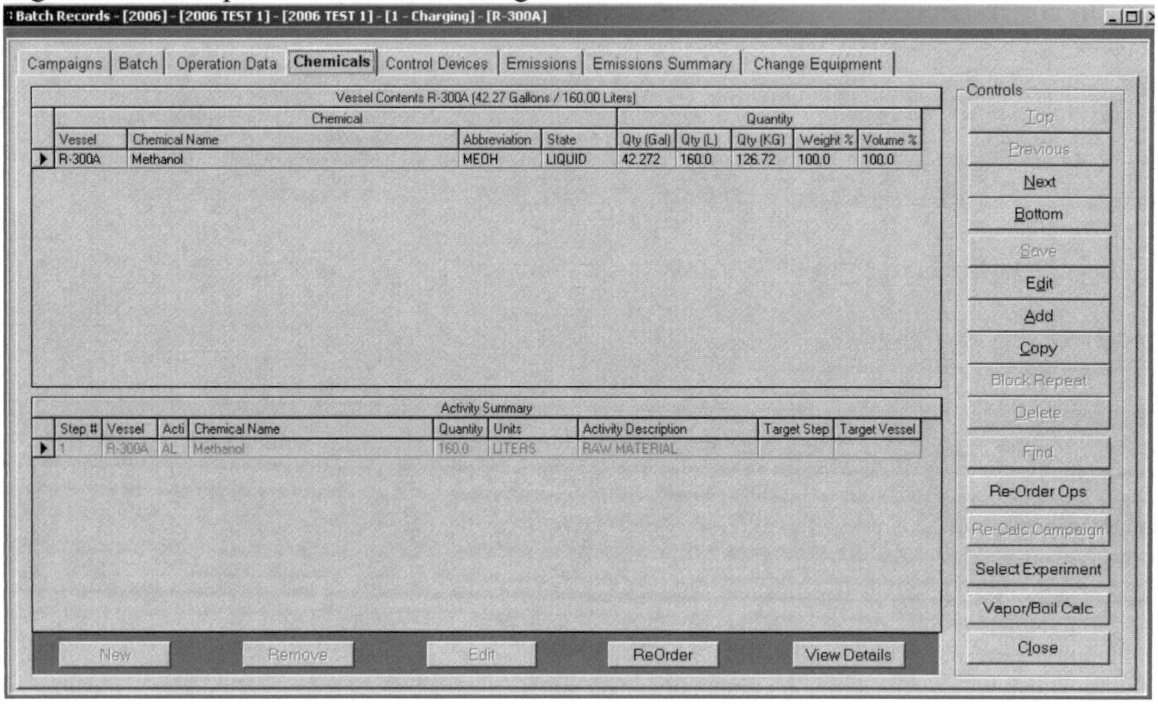

To add or remove a chemical in the Activity Summary section the buttons on the lower half of the screen are used. Selecting New, Remove or Edit displays the chemical activity screen (Figure 9).

Figure 9. Chemical Activity Screen.

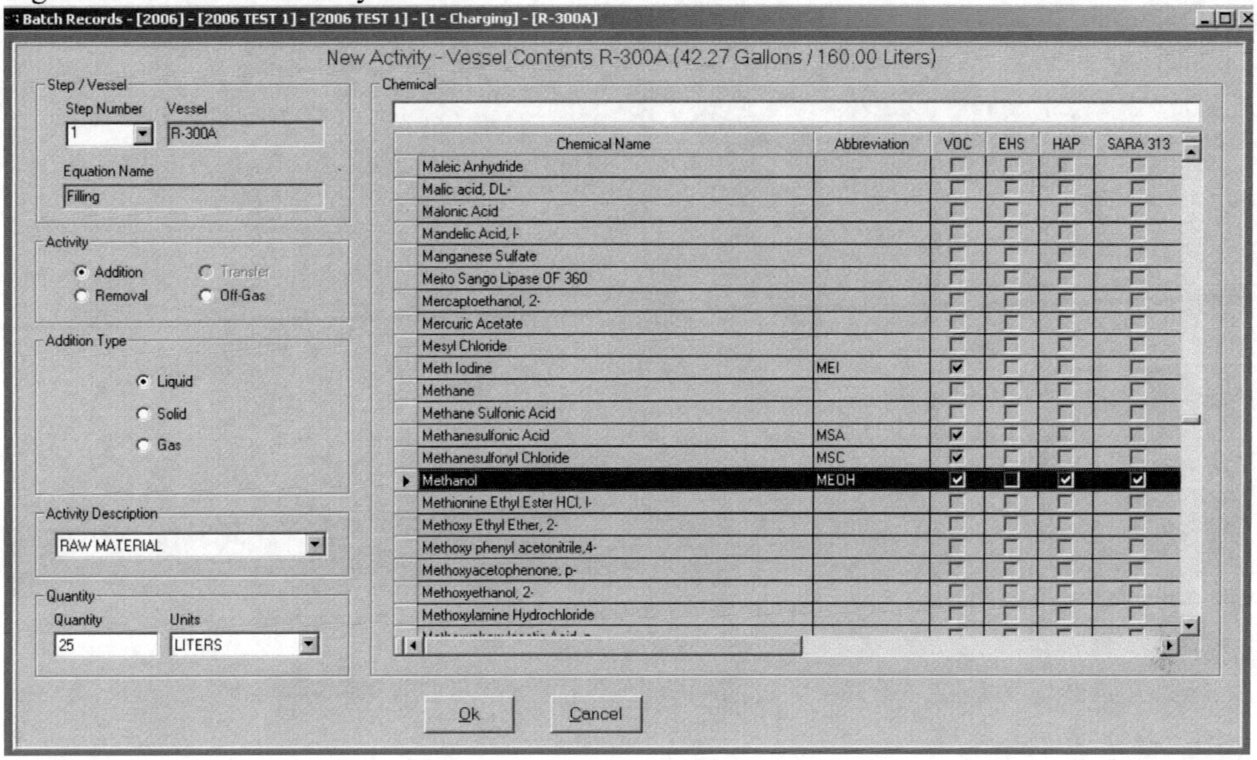

On this screen, the user can add or remove chemicals from the unit operation and set the quantity of the material selected. The addition activity allows the user to select Raw Material, Adjustment (addition) and Transfer To.

The Removal activity is where the user documents what is removed in this batch step and the fate of the removed material. The removal selections are as follows:

- Adjustment (removal)
- Dissolved Solid
- Drums
- Hazardous Waste
- Off-Gas
- Product
- Reclaim
- Transfer From

Selecting the Control Devices tab (Figure 10) displays the control devices that may be attached to the vessel in this unit operation. The user can add as many control devices as required. Clicking on a control device (highlighted) in the top grid will display the control efficiency of the device and the after control emissions for each chemical in the unit operation in the grid on the bottom half of the screen.

Figure 10. Unit Operation Control Devices screen.

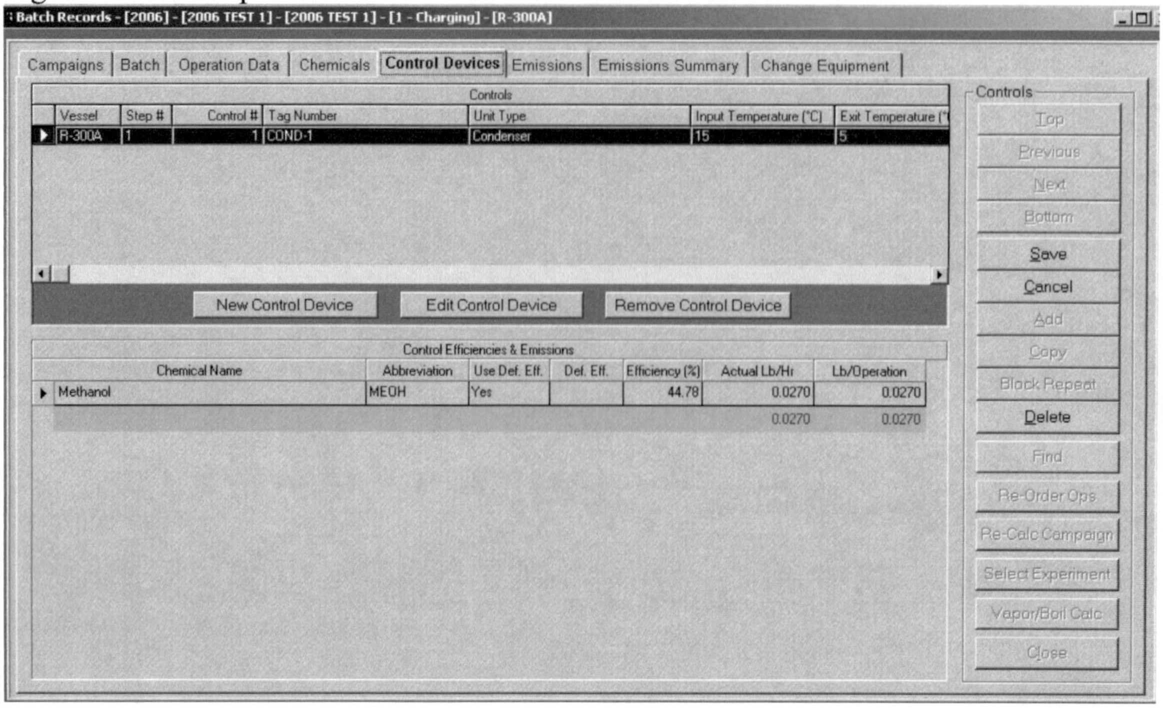

Selecting the Emissions tab (Figure 11) displays the pre- and post-controlled emission for each chemical in the unit operation in pounds per hour and for pounds per operation.

Figure 11. Unit Operation Emissions screen.

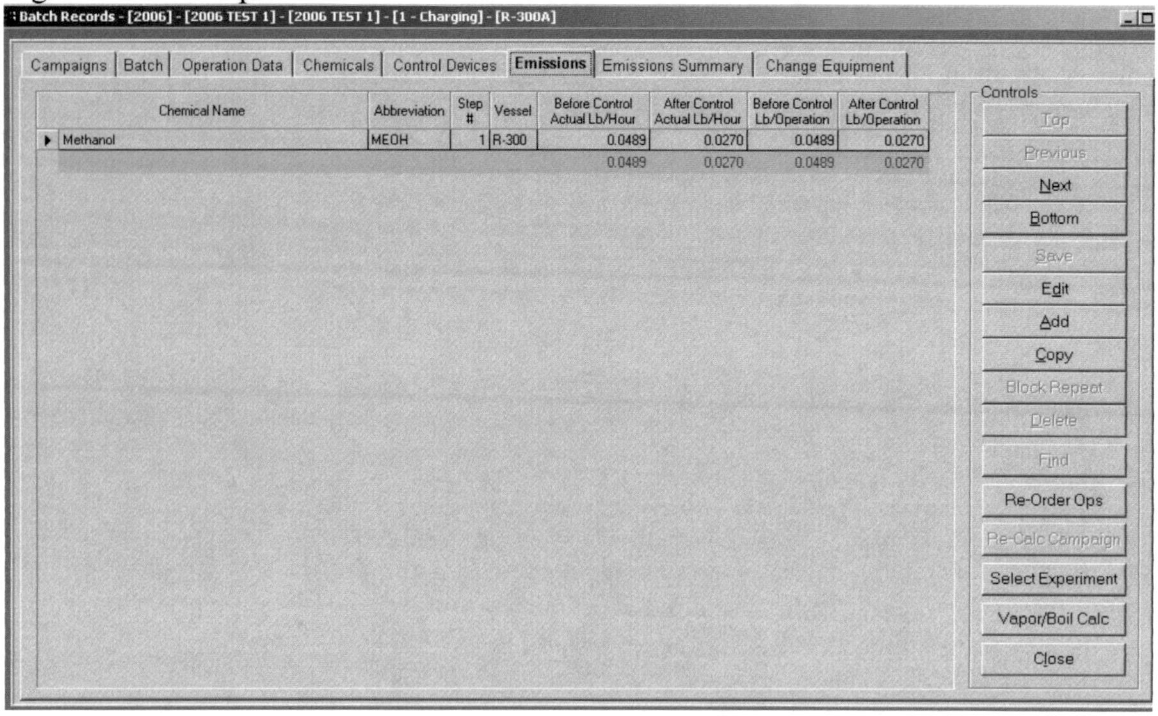

The Emission Summary screen (Figure 12) shows a roll-up of the emissions by species for each vessel in the experiment and also displays the vessel's emission limits by species.

Figure 12. Unit Operation Emissions Summary Screen.

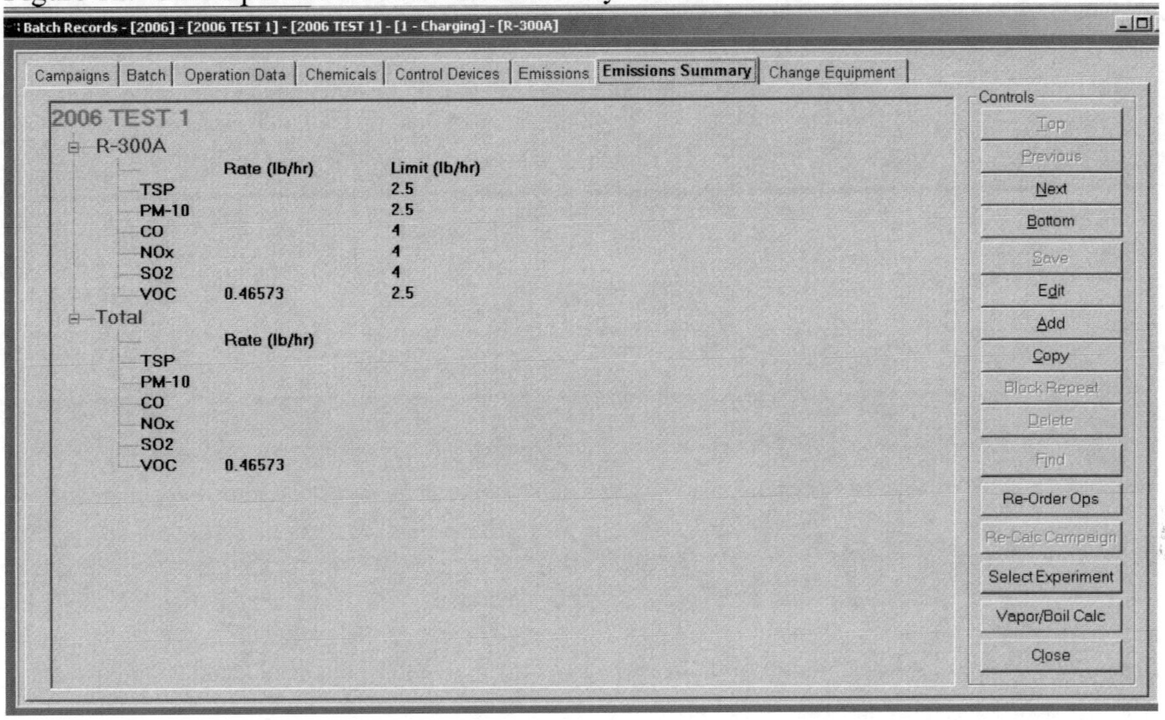

REPORTS

The reports in the Emission Profiler system are used to provide compliance information for federal, state and internal reporting needs. The various reports generated include NJDEP Post-Experiment Review, General Permit Documentation, Emission Statement, Emissions By Stack, HAPS, Greenhouse, and others, as required by the user. The reports in EP were developed using Crystal Decision's Crystal Reports.

Figure 13 shows the emissions from each vessel used in the experiment and compares each by regulated species to the required emission limits. This report is required after every experiment is run to document the experiment compliance status.

Figure 13. Excerpt from New Jersey Post Experiment Review Form.

DEQ-117F Page 2

Experiment Identification #: 2006 TEST 1

Experiment Title: 2006 TEST 1

Date: Thursday January 25, 2007 at 10:25:41PM

Source Designation:

R-300A

	LBS/HR	
Category Subcategory	**Actual Maximum**	**Permit Allowable**
CO		4.000
NOx		4.000
SO2		4.000
TSP		2.500
PM2.5		N/A
PM10		2.500
HAP	0.466	N/A
Total VOC	0.466	2.500
TXS		0.000
MeCl2		2.500
Greenhouse		N/A

☒ **Batch has met NJDEP guidelines for summary emissions limits.**

☐ **Batch has failed to meet NJDEP guidelines for summary emissions limits:**

Figure 14 is an example of one page of a Calculation Documentation for an evacuation unit operation in the batch record. Pages of this report detail the calculation methodologies and provides the specific EPA reference for the equations used.

Figure 14. Example of the Calculation Documentation Report Detailing Unit Operation Calculations.

Department: 2006	Product: 2006 TEST 1
Campaign #: 2006 TEST 1	Operation #: 4.00
Unit Type: Reactor	Unit Operation: Evacuation
Tag #: R-300A	Step #: 1
Process Condenser: None	Equation Name: Evacuation
Continuous Sweep: No Sweep Rate: 5.00	Equation Set: OAQPS 1994

Step 11. Calculate the uncontrolled emission rate of each constituent **OAQPS 1978 EQ. #2**

$$SE = \frac{(760)(X)(VRS)(MW)}{(R)(T)}$$

Where,
X = mole fraction of constituent
VRS = volume of total VOC vapor discharged, in ft^3/hr
MW = molecular weight
T = temperature (°K)
R = ideal gas law constant - 999 mmHg cubic ft/lb-mole °K

Chemical Name	Abbreviation	P (mm Hg)	X	VRS	MW	T (°K)	SE (lb/hr)
Methanol	MEOH	0.00	1.00	37.37	32.00	298.15	3.9028

Step 12. Calculate the uncontrolled emission rate per operation over multiple purges.

SE (lb/hr) = SE (lb/hr) * Number of Purges

SUMMARY

Emission Profiler is a valuable program that provides succinct environmental assessment of R&D API processes. This tool is invaluable in the lab to kilo/ pilot scale-up phase of development and is of necessity before tech transfer to a manufacturing facility is considered. Simplicity, flexibility and speed give Emission Profiler an advantage over mass-marketed alternatives that tend to create bottlenecks. Based upon sound, pre-loaded equations (OAQPS [1], CTG [2] and MACT [3]), engineers can model R&D processes and gain immediate feedback regarding environmental compliance with NJDEP-issued Air Permits.

REFERENCES

1. Control of Volatile Organic Emission from Manufacture of Synthesized Pharmaceutical Products; U.S. Environmental Protection Agency; Research Triangle Park, 1978; EPA-450/2-78-029.

2. Control of Volatile Organic Compound Emissions from Batch Processes – Alternative Control Techniques Information Document; U.S. Environmental Protection Agency; Research Triangle Park, 1994; EPA-450/R-94-020.

3. National Emission Standards for Hazardous Air Pollutants for Pharmaceuticals Production – Proposed Rule; Federal Register: April 10, 2000; 65 FR 19152.

Best Practices in the Design and Fabrication of Pilot and Demonstration Plants

David Beckman, Zeton Inc., 740 Oval Court, Burlington, ON L7L 6A9, Canada

Zeton Inc. has specialized in the design and fabrication of pilot plants and scale-up of process technology for over twenty years. Zeton has developed many best practices related to the design and fabrication of pilot plants. These include unique project execution and management approaches that save costs and shave months off project schedules, and safety systems appropriate for pilot scale plants.

This paper will discuss these methods for pilot plant design and fabrication and relate them to the changes in the scale and the demands on pilot and demonstration plants that have occurred in recent years.

Web Page: www.zeton.com

Are You Feeling Lucky? Stop Relying on Luck to Optimize Your Pilot Processes

Abstract ID# 80543

William D. Kappele
Objective Design of Experiments
Bellingham, Washington, USA

Abstract

Luck plays a surprising part in your process optimization. One-Factor-At-A-Time methods rely on luck, often producing neither the results you need nor a "map" of the process beyond the actual experiments performed. Design of Experiments eliminates the element of luck and is proven to use minimum resources to extract the maximum amount of information from a pilot plant optimization study. Response Surface Methodology enables you to understand the process at any level(s) of the factors studied. Three examples from actual pilot plant studies are presented to illustrate these points.

Introduction

Three purposes of pilot plant studies are to demonstrate viability of a new process, generate design data, and produce market development samples. At the same time the pressure is on to get to market quickly with a robust manufacturing plan. If you are lucky, everything will fall into place and you will accomplish all three goals ahead of schedule. Of course, that kind of luck is very rare. You need experimental techniques that do not rely on luck – techniques that will produce results on time even on your unluckiest day.

Design of Experiments and Response Surface Methodology are these techniques. You will see how the One-Factor-At-A-Time technique forces you to rely on luck. You will then see how Design of Experiments and Response Surface Methodology correct the weaknesses in One-Factor-At-A-Time, freeing you from dependence on luck. Finally, you will see three examples of Design of Experiments in actual pilot plant studies.

One-Factor-At-A-Time

The most popular experimental methodology is called One-Factor-At-A-Time. When using One-Factor-At-A-Time, an experimenter holds all experimental factors constant except one. This one factor is varied to discover its best level. Then this factor is held constant, and another factor is chosen to vary. This is repeated until the best level for all factors has been found. If all goes well, this will be your Sweet Spot – the combination of factor levels that maximizes product quality.

Unfortunately, it is very common to reach this point and not have found the Sweet Spot. (We will discuss why this is shortly.) Frequently the experimenter must start the process again beginning with the best combination found so far. This "starting over" often must also be repeated in an iterative process until a real Sweet Spot is found.

Weaknesses of One-Factor-At-A-Time

Why doesn't One-Factor-At-A-Time find the Sweet Spot the first time? Because One-Factor-At-A-Time has two major weaknesses:

1. One-Factor-At-A-Time does not thoroughly examine all of the possible factor combinations.

2. One-Factor-At-A-Time prevents you from studying interactions among the factors.

Let's take a look at why One-Factor-At-A-Time doesn't thoroughly cover the possible factor combinations. To do this, it helps to have a way to "see" what is happening in an experiment. A box, or cube, provides a nice picture for what can happen in an experiment. If you have three factors to study, each dimension of the box can represent the possible levels for one of the factors as shown in Figure 1. Each corner is a combination of

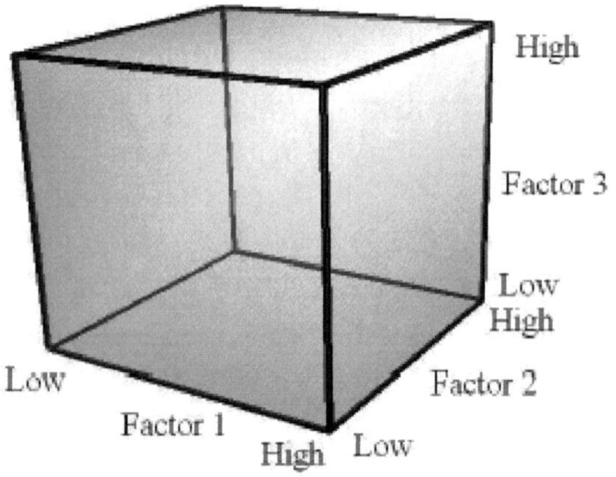

Figure 1 -- "Seeing" an Experiment

high and low levels. Each point in the cube and on the surface of the cube represents an experimental trial.

Now suppose we want to "see" what a One-Factor-At-A-Time experiment looks like. We can draw arrows on the cube representing each trial as in Figure 2. The numbers at each corner represent a hypothetical experimental response (measurement of a desired property),

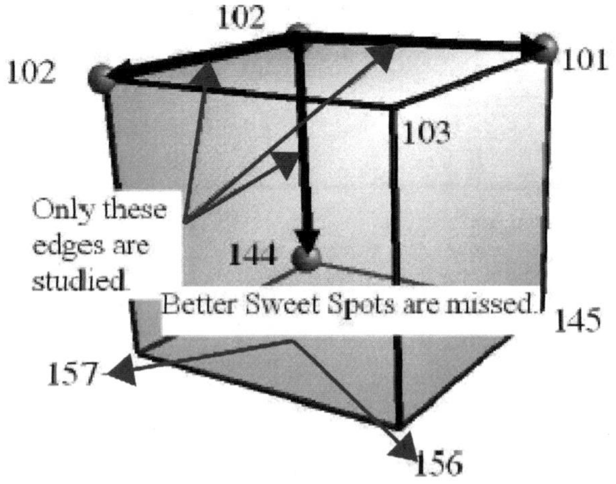

Figure 2 – One-Factor-At-A-Time

higher numbers indicating superior quality.

In Figure 2, only three edges are actually studied. The rest of the cube is left unexamined. How did this come about?

The first experimental trial was at the corner representing low factor 1, high factor 2, and high factor 3. The quality at this corner was 102. The next trial held factors 2 and 3 constant, and varied factor 1 to its high level. The quality here was 101, a decrease, so the conclusion is that factor 1 should be set low. Next factors 1 and 3 were held constant, and factor 2 was varied from high to low. It appears that factor 2 has no influence, so it was left high. Last, factors 1 and 2 were held constant while factor 3 was varied from high to low. Quality jumped to 144, leading to the conclusion that the Sweet Spot is at factor 1 low, factor 2 high (or it doesn't matter) and factor 3 low. As you can see, most of the cube remains unexamined, including two corners with significantly higher quality.

Looking at Figure 2 you can see that many portions of the cube are left unexplored. Even if a Sweet Spot is found, a better Sweet Spot, i.e. a higher quality product, could be lurking in the unexplored region. *You need luck to find the Sweet Spot!*

To see the second problem with One-Factor-At-A-Time, take a look at Figure 2 again. This time notice that when factor 3 is high (the top of the cube), factor 2 has little effect. Notice also that when factor 3 is low, factor 2 has an important effect. This is an interaction between factors 2 and 3.

Interactions are found in two ways:

1. accidentally, or

2. by varying factors together to see the effect.

One-Factor-At-A-Time, by its very *nature*,

varies only one factor at a time. **The *only* way to find an interaction with One-Factor-At-A-Time is accidentally – *by luck!***

Design of Experiments

Both of the weaknesses of One-Factor-At-A-Time are eliminated by using a designed experiment. The purpose of Design of Experiments is to acquire the maximum amount of information from a minimum amount of experimentation. By carefully considering the experimental trials to run to gain the most information ("designing" an experiment), you can not only save time and money, but also eliminate the need for luck. You thoroughly cover the cube, so that you won't miss the real Sweet Spot, and you will be able to determine the effects of factor interactions.

Run Order	%Cyan	%Yellow	%Magenta
1	3.06	2.94	2.84
2	3.47	5	5
3	5	3.39	3.67
4	5	1	5
5	2.91	3.09	1
6	1	5	3.33
7	5	5	1.23
8	1	1	1.23
9	2.61	1	3.67
10	5	1	1
11	1	2.53	5
12	1	5	1
13	3.06	2.94	2.84
14	5	3.39	3.67
15	1	5	3.33
16	2.61	1	3.67
17	1	5	1

Table 1 – An Experiment Design (Sloane) – The goal is to optimize %Dye in each of three inks between 1% and 5%to produce the largest number of printed colors

An experiment design provides a list of experimental trials, detailing the factor levels – specific values for factors, like "3.06% cyan dye, 2.94% yellow dye, and 2.84% magenta dye." You can see an example in Table 1. This is shown on a cube in Figure 3. This design is "I-Optimal," a highly efficient type of design. It spreads trials around the cube and collects the information necessary to model everywhere in the cube.

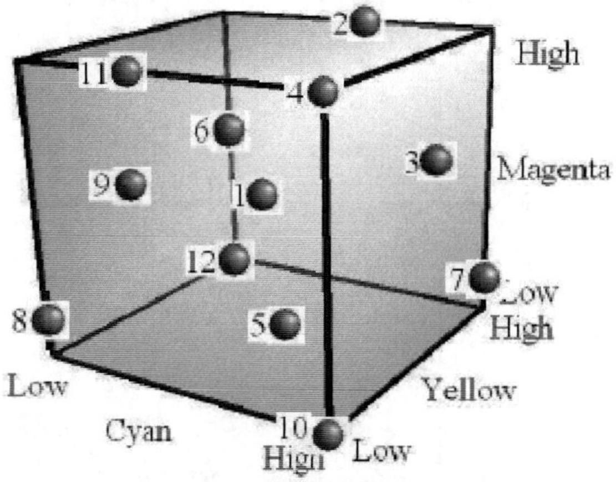

Figure 3 – An I-Optimal Design (numbers correspond to Run Order in Table 1)

The data collected using an experiment design are analyzed using Response Surface Methodology.

Response Surface Methodology

The purpose of Response Surface Methodology is to create a mathematical model using data collected with your designed experiment. With this model you can ***predict the results of experiments you have never run***, saving time and money. This model will allow you to ***predict the best conditions for meeting your goal*** – your Sweet Spot.

Suppose you are optimizing a pilot plant to produce ink-jet inks. You measure such features as stability, number of colors that can be printed, viscosity, etc. You can specify the acceptable limits for each quality feature and let your computer (using the mathematical model from your data) find the best combination of factors to meet your goals. You can even make a contour plot – a map – of how your product quality features vary as your experimental conditions vary. You can see an example in Figure 4. **With Design of**

Experiments and Response Surface Methodology you are poised to optimize your pilot plant processes *without relying on luck.*

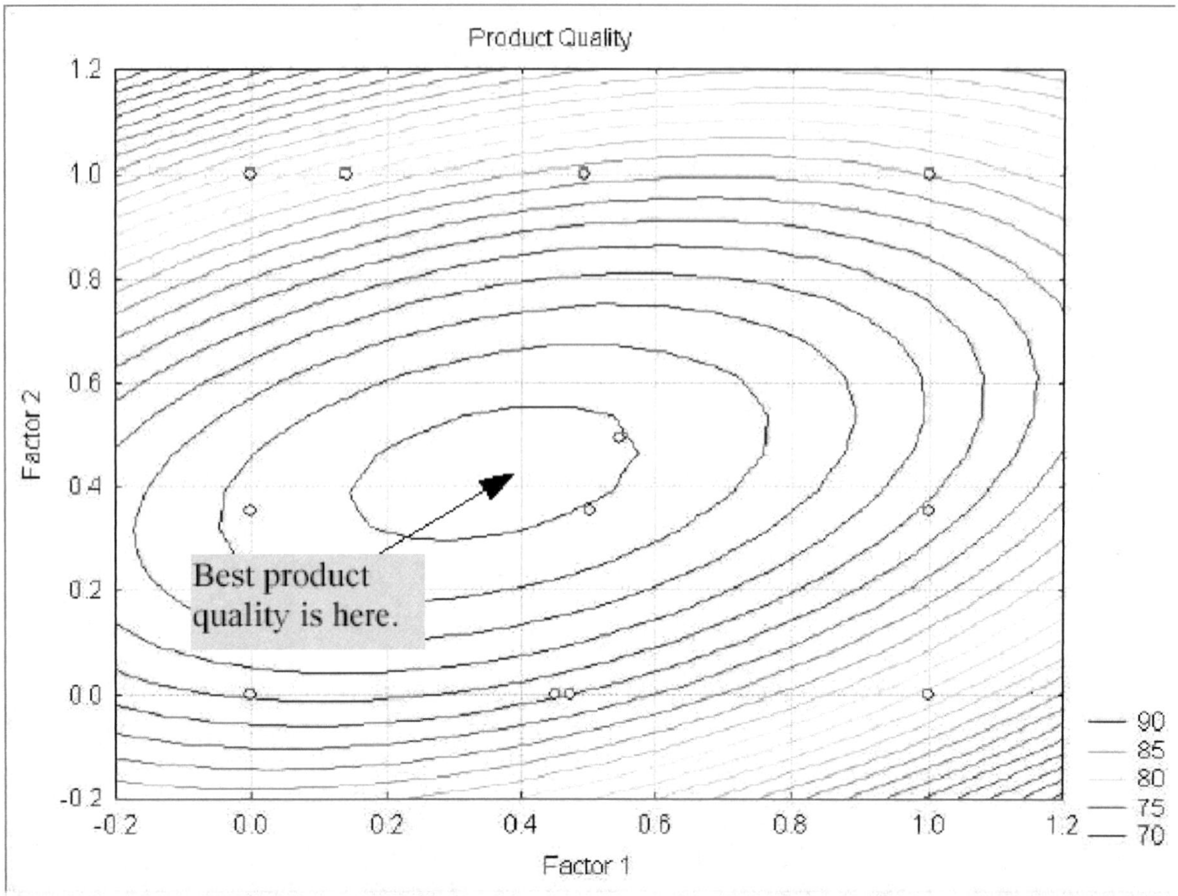

Figure 4 – A Contour Plot of Product Quality (Created by STATISTICA, www.StatSoft.com)

Example 1: Herbicide Pilot Plant

David A. Long (Long), Principal Scientist for Aventis CropScience, reported the use of Design of Experiments to optimize a pilot plant and scale it up for production of an herbicide.

The initial pilot plant optimization and scale up were performed *without* using Design of Experiments. Unfortunately, luck was not on Aventis's side – the production herbicide was clogging the spray application equipment, a result not seen with the pilot runs.

Designed experiments were applied to the pilot plant in an attempt to solve this problem. A model was created using Response Surface Methodology that predicted optimal conditions for herbicide dispersion, eliminating clogging for pilot runs. During scale-up, however, the clogging returned. It was discovered that the extrusion temperature was much higher in production than it had been on the pilot equipment, and this seemed to explain the difference.

The researchers decided to dilute the herbicide for extrusion. This was exactly contrary to the predictions of their model. In other words, they doubted the results of their designed experiment. This dilution led to exactly the instability the model predicted, confirming the accuracy of their Design of Experiments work, but leaving them scratching their heads.

Aventis's researchers began searching

for factors they had not yet studied. They discovered that kneading was a very important factor that they had not yet considered. After optimizing kneading along with the other factors, the solution was found.

David Long wrote, "The [Design of Experiments] approach was absolutely essential to determine which factors influenced the dispersion properties during storage." He further wrote, "In this case there was simply too much to risk by conducting [One-Factor-At-A-Time]."

This example not only shows the advantage to using Design of Experiments, but shows how **it combines with your knowledge of your process to solve complex problems**. Finding that kneading was important resulted from the use of engineering expertise to explain differences between predicted behavior (based on the experiment design) and observed behavior.

Example 2: Dairy Pilot Plant

C. Guillaume (Guillaume) and his co-workers at the University of Montpellier optimized a pilot plant for supplementing skim milk with calcium and phosphate.

A designed experiment was used and models were created using Response Surface Methodology. An important interaction was discovered between the amount of calcium added and the amount of phosphate added which influenced the quality of the mineral supplemented milk. At low levels of phosphate, calcium supplementation in colloidal protien was highly dependent on the amount of calcium added, while at high levels of phosphate, calcium supplementation was independent of the amount of calcium added.

The researchers studied four measures of the quality of the supplemented milk and were able to balance the requirements for all four using their models. In their words, "By using [Response Surface Methodology], we were able to conclude that a calcium supplementation of about 17.5 mmol/kg of milk favored the increase in the concentration of [the four desired] micellar constituents... ."

Example 3: Chemical Pilot Plant

Box, Hunter, and Hunter (Box) reported on a chemical pilot plant study in their textbook, *Statistics for Experimenters*. The pilot plant study is simplified in the text, but it illustrates the use of Design of Experiments and Response Surface Methodology extremely well.

The pilot plant needs to have temperature, concentration, and catalyst optimized. Temperature and concentration are continuous factors, but catalyst is not. Two different batches of catalyst were studied. While the catalysts were purchased from two different suppliers, they were supposed to be identical.

Response Surface Methodology was used to create a model of Yield based on data collected using Design of Experiments. A very surprising and important interaction between temperature and batch of catalyst was discovered – the effect of temperature on yield was much smaller with the catalyst from vendor A than it was with the catalyst from vendor B. While the catalysts were supposed to be identical, they clearly were not. Also, the yield using catalyst from vendor B was the highest ever seen. This information led researchers to consider the role and selection of catalyst much more carefully.

Conclusion

In this paper you have seen that Design of Experiments and Response Surface Methodology can be used to optimize pilot plant processes without relying on luck. You can learn more about Design of Experiments at
www.ObjectiveDOE.com

References

Box, Hunter, and Hunter, *Statistics for Experimenters*, J. Wiley and Sons, 1978, ISBN 0-471-09315-7, pps. 307-319.

Guillaume, et al, "Effect of Salt Addition on the Micellar Composition of Milk Subjected to pH Reversible CO2 Acidification," *J. Dairy Sci.* 85:2098-2105, pps. 2098-2105.

Long, David A, "Pilot-to-Production Requires Design of Experiments to Unclog Herbicide," http://www.statease.com/pubs/herbicide.pdf

Sloane, NJA and R.H. Hardin, Gosset: A general-purpose program for designing experiments,
www.research.att.com/~njas/gosset/.
The design in Table 1 was created using Gosset. It is available in the I-Optimal Design Assistant at
www.objectivedoe.com/Assistant/assistantintro.php

Looking for More?

Here are some interesting references for further information:

"Design of Experiments with Bill Kappele," a free DVD discussing the advantages to using Design of Experiments and Response Surface Methodology, available from Objective Design of Experiments,
http://www.objectivedoe.com/DVDrequest.php

Dragone, et al, "Improvement of the Ethanol Productivity in a High Gravity Brewing at Pilot Plant Scale," *Biotechnol Lett.* 2003, July 25(14), pps. 1171-1174.

Lee, et al, "TVA Pilot Plant Phosphates Flotation Conditioning Study," *Nutrient Cycling in Agroecosystems,* Vol 29, No 3, Sept. 1991, pps. 347-363.

Canettieri, et al, "Optimization of Acid Hydrolysis from the Hemicellulosic Fraction of Eucalyptus Grandis Residue Using Response Surface Methodology," *Bioresour Technol,* 2007, Jan, 98(2), pps. 422-428.

Ahmad, et al, "Optimization of Coagulation-Flocculation Process for Palm Oil Mill Effluent Using Response Surface Methodology," *Environ Sci Technol,* 2005, April 15, 39(8), pps. 2828-2834.

Kappele, William D, "Quantifying Color Gamut," Proceedings of IS&T's NIP 13: International Conference on Printing Technologies, Nov. 2-7, 1997, pps. 470-474. (This paper includes an example using Design of Experiments and Response Surface Methodology.)

Pilot Plants: Best Practices Benchmarking and Open Forum Discussion

Joseph B. Powell, Chemical Process Technology, Shell Global Solutions US Inc., Westhollow Technology Center, P. O. Box 1380, Houston, TX 77251-1380

Highlights of the AIChE 12B Pilot Plants Benchmarking Survey (2003/4) will be presented, along with open forum discussion of best practices and issues in the implementation of pilot plant programs.

In Situ Mid-Infrared Spectroscopy as a Tool in the Development of a Continuous Process for the Production of Cyclopropylcarbonitrile

Norma L. Buchanan, Technology Department, Eastman Chemical Company, P.O. Box 1972, Kingsport, TN 37662-5150

Mid-infrared in situ spectroscopy is a valuable research and manufacturing tool that has been employed at Eastman Chemical Company for many years. As an example of the use of this technology, piloting and production of cyclopropylcarbonitrile in a short-term manufacturing campaign will be presented. Cyclopropylcarbonitrile had previously been produced by batch reaction of hydroxyl amine with cyclopropylcarboxyaldehyde; both hydroxyl amine and an expected oxime intermediate are known to decompose with high energy release. In developing a continuous process, it was imperative that the reactor concentrations of hydroxylamine and the oxime intermediate be maintained at appropriately low levels. The use of mid-infrared in situ spectroscopy enabled safe operation during the piloting and subsequent production of cyclopropylcarbonitrile.

Real Time Measurement of Propylene Oxide in Polyether Polyols Via Calorimetry

Paul Weider, Process Development, Shell Global Solutions US Inc., Westhollow Technology Center, 3333 Highway 6 South, Houston, TX 77082

Shell Chemicals is among the leading global producers of polyether polyols. CARADOL polyether polyols, when combined with isocyanates, are used in urethane applications. Urethanes find use in products such as flexible and rigid foams, and in Coatings, Adhesives, Sealants & Elastomer (CASE) systems. As a result we may encounter them in a wide variety of goods including furniture, car seating, bedding, paints and coatings, artificial sports tracks, playground surfaces, ski suits and other waterproof leisure wear. Shell produces more than four hundred thousand metric tons per year of polyether polyols.

Polyether polyols are produced by base or acid catalyzed polymerization of propylene Oxide (PO) with polyhydric alcohols and/or other functional initiators. An important aspect of safeguarding these exothermic reactions is to maintain a safe solution concentration of PO. Knowing exactly how much PO is in solution is difficult to ascertain due to the physical properties of the reacting mixture. Grab samples are hazardous to collect and difficult to analyze. In situ Infrared (IR) spectroscopy can readily detect the PO in solution, but how to calibrate the resulting IR spectra? We have accomplished this through reaction calorimetry. By adding pulses of PO to the reacting mixture and back integration across the pulse of the heat released, we can estimate the concentration of PO at any given time in the reaction. This is matched to the corresponding IR spectra to create calibration sets for the various grades. This work will also discuss moving these calibrations across platforms with different probes and detectors.

Ethylene Oxide Hydroformylation in Expanded Solvents

Joseph B. Powell and Paul R. Weider
Process Development
Shell Global Solutions U.S. Inc.
Westhollow Technology Center
P.O. Box 1380, Houston, TX 77251

Introduction

Hydroformylation of ethylene oxide (EO) is an efficient route to the production of 1,3-propanediol (PDO), commercialized by Shell Chemicals in 1996. Primary use of PDO is in production of polytrimethyleneterephthalate (PTT), a new polymer with advantaged properties and reduced environmental footprint relative to traditional nylon. Use of supercritical solvents or solvents "expanded" via addition of a near- or super-critical component has attracted recent attention for homogeneously catalyzed reactions.[1-3] The purpose of the current study was to assess potential of near-critical expanded solvents for hydroformylation of ethylene oxide with homogeneous cobalt carbonyl as catalyst.

The study is an example of use of off-line or bench scale methods in process screening. By careful design of experiments and judicious use of analytical tools (*in situ* Fourier transform infrared spectroscopy), commercial potential can often be assessed without having to resort to a full-scale continuous pilot plant.

Expanded Solvent Options and Experimental Apparatus

Reaction conditions for ethylene oxide hydroformylation entail temperatures between 60 and 120 °C, and pressures of 50 – 200 bar (750 – 3000 psi). Potential near-critical expansion co-solvents under these conditions can be identified from published literature.[4] In addition to offering critical points near those of reaction conditions, the co-solvent must also be inert towards cobalt carbonyl hydroformylation catalyst, as well as EO and hydroformylation product 3-hydroxypropanal (HPA):

$$EO + CO + H_2 \rightarrow 3\text{-HPA}$$

Moreover, the resulting solvent blend must favor the formation of 3-HPA, vs. rearrangement or isomerization of EO to acetaldehyde. A key feature of co-solvent candidates is that partial pressure at the critical point takes up a substantial portion of the design pressure for EO hydroformylation (50 – 200 bar design).[4]

Hydroformylation reaction and fluid mixing tests were conducted in a 300-ml autoclave equipped with a hollow-shaft mixer for gas re-dispersion. After mixing, contents could be pressured into a sight glass to measure volume of liquid phase with dissolved near-critical co-solvent phase. Alternately, solvent expansion could be assessed by prefilling the sight glass with liquid solvent, followed by slowly bubbling near-critical solvent into the mixture from the bottom, with overnight equilibration. An FTIR with *in situ* CIR cell was used to assess carbon monoxide (CO) solubility as a function of solvent blend, in separate experiments.

Results of Expanded-Solvent Tests

Results of near-critical co-solvent expansion tests conducted via sight glass measurement are shown in Table 1. Thermal expansion of solvent A was taken as base case, and various near-critical co-solvents were added at the indicated pressure ("NC-psi") to expand the liquid, as measured by sight-glass level. Relative syngas solubility in the liquid phase was estimated as the relative liquid-phase expansion, divided by the fraction of the total pressure (100 bar) occupied by syngas, after consideration of the partial pressure required to support the near-critical solvent component. This model assumes simple gas solubility via occupation of cages within the liquid structure, in the absence of chemical interaction, in direct proportion to the partial pressure (Henry's law). For experiments with near critical co-solvent above the critical point (CO_2, ethane), the solvent expansion was notable (11 – 21%). However, excessive partial pressure was consumed in supporting the near-critical component, such that allowed syngas partial pressure at 100-bar total pressure, was reduced substantially. Use of near-critical co-solvent below the critical point (propane, dimethylether (DME)) produced less expansion but consumed less of the available system pressure.

Table 1: Near-Critical Co-Solvent Expansion Tests

100-bar total pressure (NC = near critical solvent + 2:1 syngas)

Solvent	NC-comp	Reaction T deg C	NC-comp Tc deg C	T-Tc deg C	% expand liquid	NC-comp partial press. bar	Relative syngas Solubility
Solvent-A	none	80	na	|x| > 120	5%	0.10	1.00
Solvent-A	propane	80	96.67	-16.7	8%	12.58	0.91
Solvent-A	CO2	80	31.3	48.7	21%	52.78	0.56
Solvent-B	ethane	67	32.28	34.7	11%	53.01	0.51
Solvent-B	ethane	77	32.28	44.7	18%	54.57	0.52
none	DME	80	126.9	-46.9	11%	7.85	0.98

DME = dimethyl ether

Tc = critical temperature

Table 2 shows EO hydroformylation studies conducted in the 300-ml stirred autoclave at 100 bar total pressure, and 373 K to assess kinetics and selectivities. The program sought to minimize selectivity to acetaldehyde relative to hydroformylation product 3-HPA, as reflected by the molar yield of hydroformylation product (HPA). Use of CO_2 (run B) to expand solvent-A led to decreased overall rates (TOF) at equivalent selectivity, within experimental error. When analyzed on the basis of TOF divided by the syngas partial pressure present in the reactor, the rates per unit of syngas were indeed higher with the expanded solvent systems. This is of limited practical advantage, however, if a higher total pressure is required to achieve the same absolute rate, considering the partial pressure of the expanded solvent system itself. A similar result is obtained for runs D & E with near critical ethane vs. hexane as diluent for solvent-B. TOF per psi syngas was higher for the ethane-expanded system, though absolute rates were less than for the hexane / solvent-B system, where hexane did not back out the allowed syngas partial pressure. Run F with an excess of near-critical DME provided good molar yields, but rates were poor relative to runs A or D where near-critical solvent was not present. Thus, in no case did absolute rates with near-critical solvent or co-solvent exceed those of similar runs with solvents not near their critical point.

Table 2: EO Hydroformylation Batch Experiments
80 C, 100-bar 4:1 syngas.

Run #	Solvent	Syngas bar	NC-comp Tc deg C	T - Tc	Relative TOF	molar HF yield	TOF/ 100-bar syngas
A	Solvent-A	98.6	n.a.	n.a.	1.00	0.805	1.014
B	700-psi CO2 / Solv-A	47.6	31.3	48.7	0.73	0.760	1.527
D	60% hexane / Solv-B	98.6	n.a.	n.a.	1.09	0.770	1.104
E	500-psi ethane / Solv-B	64.6	32.28	47.72	0.81	0.750	1.254
F	80% DME/Solv-C	80.9	126.9	-46.9	0.62	0.810	0.761

Molar HF yield = yield of hydroformylation product from EO
Solvents A – C have critical temperature > 120 C above reaction temperature.

FTIR Study of Syngas (CO) Solubility

Spectra of CO dissolved in solvent “D”, a preferred solvent for hydroformylation, as a function of pressure at 80 °C is shown in Figure 1. Tests were repeated with 200 psi of CO_2 (not shown), leading to diminished absorbance of Solvent D itself. This can be attributed to possible solvent expansion as well as change in effective CIR effective path length due to change in refractive index of the composite solvent. If corrected for attenuation of SOLVENT D signal assuming all is attributed to effective path length and not physical dilution, a maximum possible CO solubility is calculated. Maximum CO solubility at a given partial pressure is observed to be less in the presence of CO_2, relative to SOLVENT D only with no CO_2 (Figure 2). CIR cell measurements thus failed to provide any conclusive evidence that CO_2 expansion led to an increase in CO solublization at a given partial pressure of syngas. This result suggests any rate enhancements with CO_2 co-solvent in Table 2 per unit syngas partial pressure may be attributed to solvent interactions, more so than enhanced syngas solubility per unit of syngas.

Figure 1: FTIR spectra for water-saturated Solvent D under nitrogen, heated to 80°C with addition of 1:1 synthesis gas (H_2 / CO) gas to 20 – 1600 psig. Gas phase carbon monoxide has the strongest absorbance in the region from 2230 to 2050 cm^{-1} (and CO_2 from 2390 to 2300 cm^{-1}).

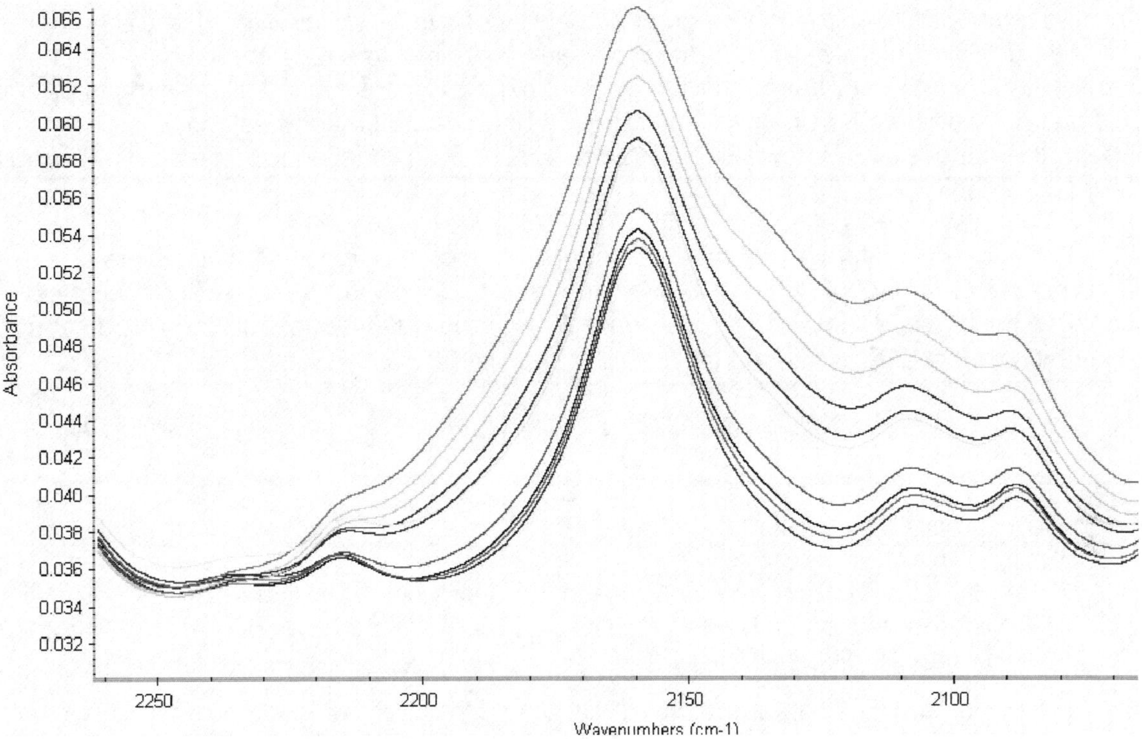

Figure 2: FTIR CO absorbance with and without 200 psi CO_2 (stabilized at 286 psig upon heating to 80°C) followed by 1:1 syngas addition to increase CO partial pressure.

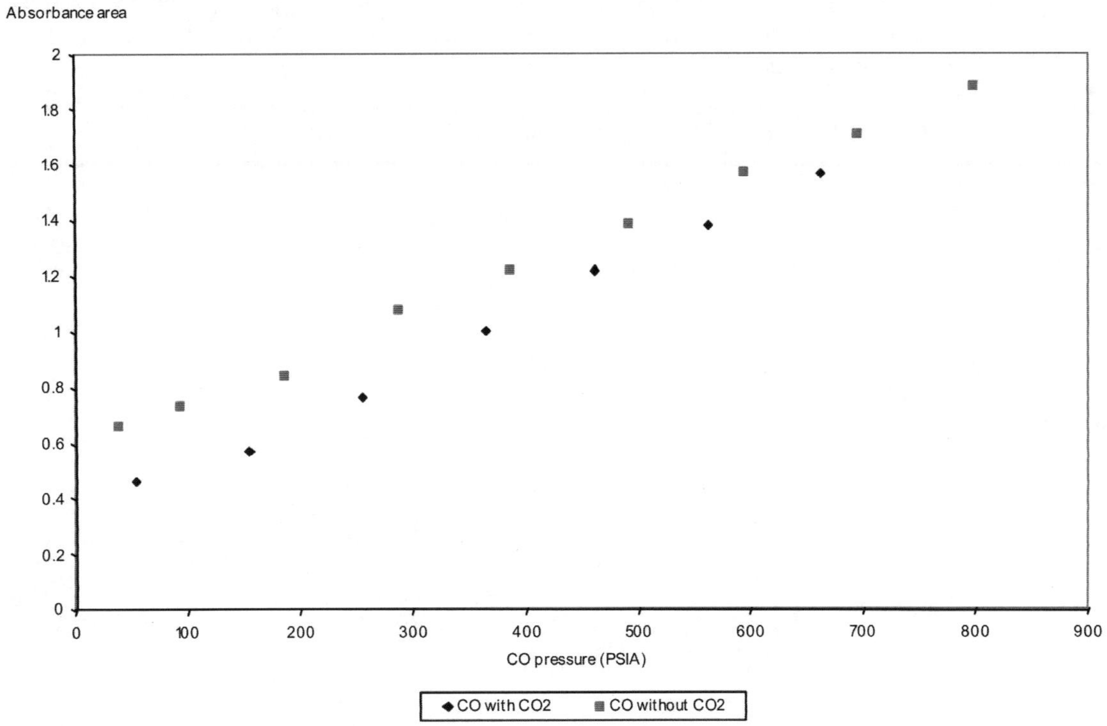

Discussion

Results of this investigation show that near-critical expanded solvents can be used in EO hydroformylation to provide reaction selectivities and rates in expanded solvents that would otherwise require higher partial pressures of syngas. However, there is no benefit in total design pressure, as the partial pressure required to supply the near-critical co-solvent offsets the enhanced performance of expanded solvents at lower syngas partial pressure. The primary process advantage in use of an expanded solvent system may be in the opportunity for flash separation to concentrate reaction products, post reaction, for some design concepts.

FTIR is a useful tool for direct measurement of carbon monoxide solubility, with correction for impact of solvent composition on path length for an internal reflectance cell. In this case, direct FTIR measurement provided no evidence for enhanced syngas solubility in the CO_2-expanded solvent D system at 80 °C.

References

1. H. Jin and B. Subramaniam, Rhodium-catalyzed Homogeneous Hydroformylation of 1-octene in CO2-expanded Solvent Media, AIChE Fall 2002 Indianapolis, Paper 235h. .
2. H. Jin and B. Subramaniam, "A Study of Rh-catalyzed hydroformylation of 1-hexene in Supercritical Carbon Dioxide", AIChE 2003 Fall National Meeting, San Francisco, CA, paper [546d].
3. C. S. Fujita, S. Fujisawa, B. M. Bhanage, Y. Ikushima, and M. Arai, Hydroformylation of 1-hexene with rhodium fluorinated phosphine complexes in supercritical carbon dioxide and in conventional organic solvents, *New. J. Chem.*, *26*, 1479-1484 (2002).
4. C. L. Yaws, Chemical Properties Handbook, McGraw-Hill, NY (1999).

Safe Use of Microwaves in Chemical Processing

By David C. Attride
Senior Development Associate
Eastman Chemical Company
Kingsport, TN

Microwave heating is an unusual technology in the sense that it is a technology far more commonly used in residential kitchens than in industrial manufacturing settings. The development of the low-cost magnetron in the 1960s led to relatively inexpensive home microwave ovens. As a result, all of us have used 1kW, 2.450 GHz microwaves to heat coffee or cook food but very few chemical engineers have used "high power" microwaves to heat flammable mixtures. The purpose of this talk is to acquaint you with idiosyncrasies of microwaves and discuss aspects of safely using microwaves with flammable materials.

What are the advantages of using microwave energy to heat materials instead of using steam, hot oil, electric resistance heaters or direct-fired furnaces? Very high energy densities can be created using microwaves so very rapid heat-up times are possible. The energy transfer from microwaves provide volumetric heating instead of conduction or convection so microwaves would be a good choice for heating objects in a vacuum and are capable of providing uniform heating. Microwaves are a nearly infinite temperature heating source so very high temperatures can be achieved (i.e. melting ceramics; pyrolysis of organics; generating plasma). Microwaves can be extremely useful for heating materials that have low thermal conductivity such as foams or wood. Microwaves have been shown to generate localized superheating of solvents which has been shown to substantially increase reaction rates. [1]

What are the potential disadvantages of using microwaves for heating? Microwaves are a relatively high cost means of providing heating. The capital cost for a high power microwave generator is 3x-6x that of providing large-scale steam generation. Robert Schiffmann has published an estimate that it costs $375,000 to $750,000 in capital costs to provide microwave generator, applicator, power transmission and control system.[2] Heating solids with microwaves can result in non-uniform heating (think of your experiences in cooking a frozen dinner). As a result manufacturers of home microwaves add a spinning chopper to deliver multi-mode microwaves and often suggest a turntable all in an effort to more uniformly heat the food. Heating solid objects with microwaves generally require the solids be free of metals objects. When sharp-edged objects are present in microwave heating chambers, severe arcing results. Plasmas can be generated by application of high microwave field density in deep vacuum applications. Non-polar solvents are difficult to heat with microwaves (this can be a disadvantage but in some cases may be used advantageously).

Lastly, while microwave ovens have been ubiquitous in kitchens for the last several decades, industrial microwave systems are still rather new in chemical processing. Microwave-assisted chemical synthesis has been an active area for research for the last 5-

10 years but most of the literature describes laboratory research systems rather than industrial-scale applications. There is a rather limited range of companies and consultants that offer industrial-scale equipment and expertise.

Many of the known industrial uses of high powered microwaves (6kW or greater) involve heating materials that contain water. In the food industry, continuous conveyor cooking in a microwave chamber is utilized for cooking bacon on a large scale. Similar technology is used for ultra rapid pasteurization and for heating high viscosity water-based materials. High power microwaves are used to generate plasma for certain aspects of semiconductor manufacturing. Some high-value ceramics are cured using microwaves. In the chemical, building material and plastics industries microwaves are used to dry materials, cure resins and provide energy for calcinations.

A little background on microwaves and microwave terminology is useful. The portion of the electromagnetic spectrum from 300 MHz to 300 GHz is called microwaves. For instance, your home microwave operates at 2,450 MHz (wavelength = 0.122 meter) and generates 1kW of microwave power. Typical industrial microwaves operate at 915 MHz (wavelength = 0.327 meter) and can generate 6kW to 100kW of power.

The microwaves are typically generated by a magnetron which is powered by 2,000 to 8,000 volts. Industrial-scale magnetrons have a sink for reflected power that is water cooled to prevent damage to the magnetron by high reflected power (caused when a chamber contains materials that have little or no microwave absorptivity). When non-flammable materials are being heated (i.e. ceramics or water-based foods, etc.), the microwave generator can be located at the point of processing. When the microwave is used to heat flammable materials, the microwave generator is located in a non-explosion proof area and the microwaves are transmitted to the processing unit (often called an applicator or launcher) via waveguides. Waveguides are typically rectangular metal conduit designed specifically for the wavelength of the microwaves used (so that a specific mode of microwave is propagated). The waveguide run can be 10s of meters without significant attenuation. Waveguides direct the microwaves to the processing point and connect to an launcher that propagates a mode of microwaves designed to most efficiently and uniformally heat the material. The launcher might be built into a pressure vessel or it might be built into a continuous conveyor heater (with special baffles to prevent microwave leakage out the open ends of the tunnel heater). It is possible to beam microwaves directly into a processing vessel by mounting the waveguide directly to a flanged barrier window (not simply a sightglass). This implementation does not allow control of the microwave mode or the distribution of energy within the processing vessel.

Even though each of us have used microwaves at the 1kW scale in our kitchen, implementation of an industrial-scale, higher power microwave may cause concern amongst the operating staff. Some education about microwave health issues and carefully describing the use of the phrase "microwave radiation" is necessary. Microwaves are not sufficiently energetic to be classified as ionizing radiation and therefore have not been shown to cause cancer. Exposure to microwaves caused localized heating of the tissue exposed. Tissues especially sensitive so localized heating are: eyes and reproductive organs. Therefore, any industrial use of microwave must take precautions to prevent microwave leakage. Potential pathways for leakage are: waveguide seams & flanges; vessel sight-glasses; vessel man-ways & flanges; vessel nozzles larger than one-half of the microwave wavelength. Unless a cutoff screen is used, microwaves can travel out of these larger nozzles. The maximum exposure standard set by the FDA is leakage of 5mW/cm^2 whereas the industrial standard is no more than 10 mW/cm^2.

Figure 1 Basic Microwave Heating System

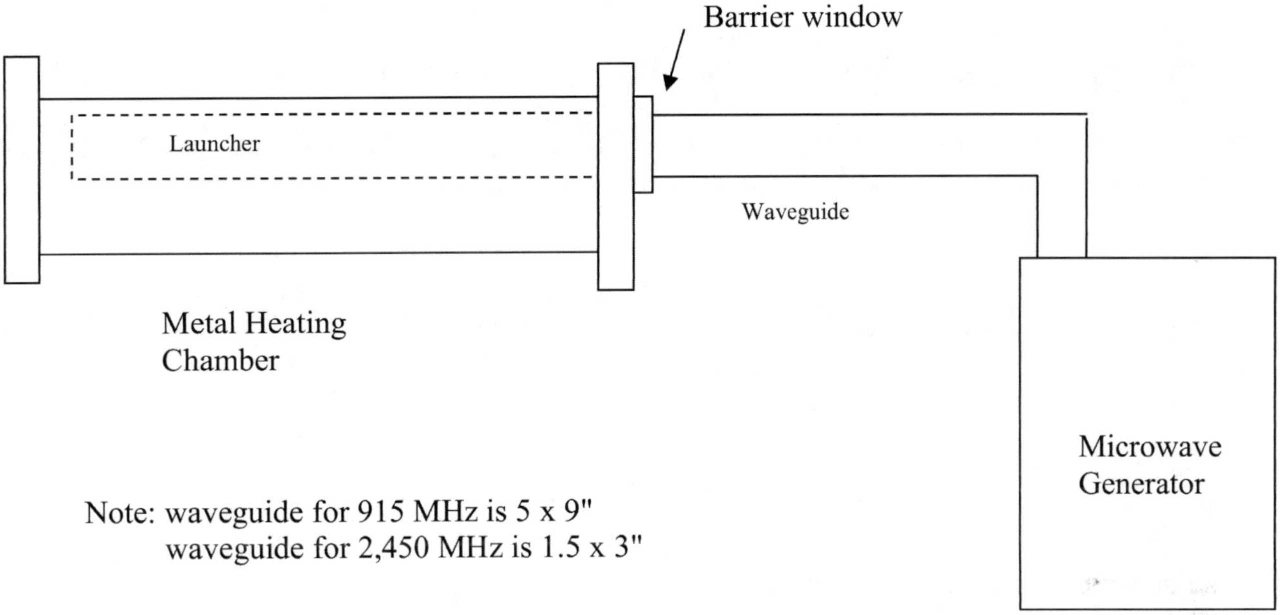

Note: waveguide for 915 MHz is 5 x 9"
waveguide for 2,450 MHz is 1.5 x 3"

This figure shows schematically a basic system for heating materials with microwaves. The microwave generator at lower right generates microwaves that exit the top of the generator and propagate within the waveguide to the heating chamber. Since the heating chamber may be evacuated, a barrier window provides a vacuum barrier and transmits the microwaves. In this example a launcher is specially designed to distribute the microwaves within the heating chamber.

Additional details regarding the microwave generator: 915 MHz units are available in 6kW, 30kW or 75kW models. 440V power is typically required which inside the generator cabinet is transformed to 2,000-8,000 volts to power the magnetron. Industrial microwave generators provide variable power generation unlike home microwaves which operate at full power but time proportionate. Instrumentation within the generator measures and displays the power generated (forward power) and the amount of microwave energy reflected by the system components back to the magnetron (reflected power). The reflected load is dissipated into a water-cooled jacket. Because of the high voltages used and cooling air vents, the microwave generator must be located in a general purpose electrical area.

The waveguide is a thin-walled, metal rectangular conduit which allows passage of the microwaves from the generator to the point of use with little attenuation. Waveguides are typically fabricated from aluminum and must be constructed with several specific microwave properties in mind. The size of the waveguide determines the mode of microwaves that are propagated. Waveguides are made in flanged sections but no gaskets are used at the flanges. Rather, metal-to-metal joints prevent microwave leakage. This requires strict control of the degree of warpage of these welded assemblies. Where necessary, waveguide bends are used but their design also requires knowledge of microwave propagation. The expense of the waveguide sections suggests the placement of the generator relatively close to the launcher/heating chamber. When objects containing flammable materials are being heating, the waveguide system should be purged with an inert gas to reduce the possibility of a fire in the event of the failure of the barrier window.

The barrier window is a specially designed microwave component that allows passage of the microwaves with minimal attenuation or reflection and seals the heating chamber so that vacuum or pressurized operation is possible. This is not simply a CPI sightglass but instead of a window made of very pure quartz or alumina or Teflon. Examples exist of unexpected microwave resonance within a barrier window leading to severe damage of the window.[3] Calculations or experiments to determine microwave reflections and/or resonance behavior are necessary for proper design and reliable operation.

Since barrier windows are brittle materials, an understanding of their pressure rating is vital to safe design when the heating chamber is a pressure vessel. Keep in mind that the suppliers of microwave components typically service industries different from the chemical industry and may have different expectations regarding equipment design and pressure rating and testing. Plans for operation at positive pressure may lead to the conclusion that a single barrier window provides insufficient protection against failures that could result in flammable materials being blown into the waveguide and possibly into the high-voltage magnetron. The use of double barrier window designs should be considered.

Even a well designed system of waveguide, barrier window and launcher will reflect some microwaves. A stub tuner is a component that is mounted close to the barrier window that reduces microwave reflection by creating destructive interferences. The stubs are brass threaded bolts mounted in a flanged waveguide section. The stub tuner is typically adjusted manually but occasionally automated.

The final component of the microwave transmission system is the launcher (or applicator). Its function is to efficiently "launch" the microwaves into the heating chamber in the mode that provides the desired heating and in a manner designed by the microwave engineer to provide uniform heating. The launcher is typically bolted directly to the barrier window and mounted to the inside of the heating chamber. One design used for a long cylindrical heating chamber utilized numerous slots along the length of the launcher. For sealing against operating pressure, each slot was covered with a flanged ceramic barrier window. In this manner the barrier window in the waveguide served as a secondary barrier against leakage of flammable chemicals. The figure below shows a schematic view of this design:

Figure 2: Launcher Schematic

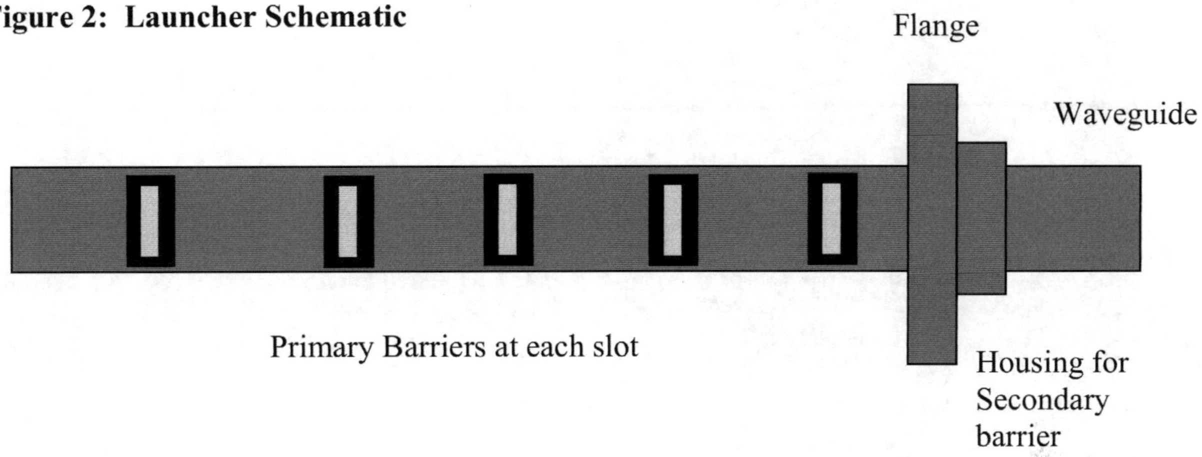

The application at Eastman Chemical was the first use of microwave heating in our pilot plant. Operators were carefully trained to make them aware of the properties of microwaves, how to detect microwave leakage and details of the design that prevented microwave leakage. A hand-held microwave leak detector was purchased and mounted in a sealed box for use in a Class 1, Div. 2 area. Vessel end flanges were specially designed for metal-to-metal sealing to prevent microwave leakage. Each flange also incorporated an o-ring groove to prevent chemical and pneumatic leakage. Sheet Teflon gaskets can not be used (unless in a recessed groove) since microwaves readily penetrate Teflon and would leak from the vessel. Flexitallic gaskets may not be useable since the close metal-to-metal gaps might cause microwave-induced arcing. To prevent microwave use during flange or sightglass removal, a pair of proximity switches was mounted at the charging flange at at the sightglass flange. Closure of all proximity switches was necessary for providing power to the microwave magnetron.

Figure 3 ETS-Lindgren Microwave Survey Meter

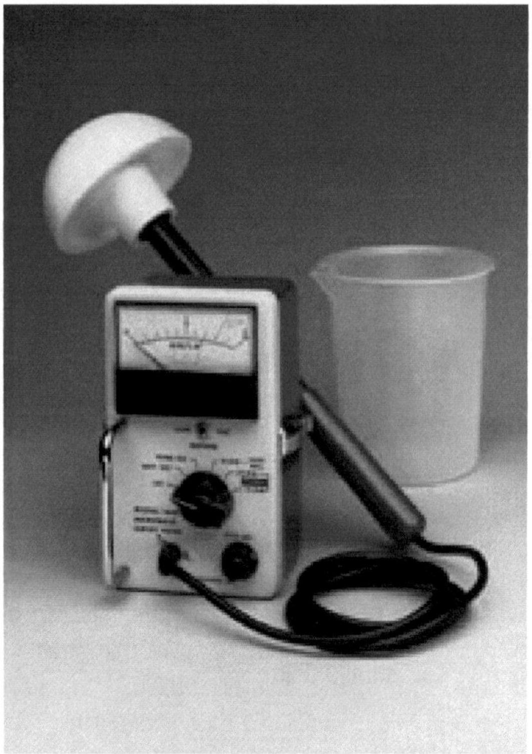

This microwave leakage meter costs about $1,500 and has a range from zero to 10mW/cm2. It is not certified for use in Class 1, Div. 2 areas so it was mounted in a sealed, plastic "Pelican" box such that the cord and sensor wand hung outside the box. It was used at the beginning of microwave heating for each run to verify that the system was microwave tight. In all our work no microwave leakage was detected so we used a nearby kitchen-type 1kW microwave oven to verify the leakage meter was working.

When heating objects with microwaves, it is vital to be able to accurately measure temperatures. In our case, the heating chamber was steam jacketed but it was important to determine the temperatures at various places in the objects during microwave heating. The old rule of thumb that you can not place metal objects in microwaves is a generalization. The launcher itself was made of 316 S/S and was mounted inside the heating chamber. Temperatures were often measured using a 316 S/S sheathed thermocouple. No arcing was ever detected but microwave heating of the thermocouple (ohmic heating) elevated the temperature of the thermocouple by 50-100 °C. Gold plating of this thermocouple reduced ohmic heating significantly. For best accurate temperature measurement, fiber optic temperature sensors were used. They were manufactured by FISO (Quebec City, Canada) and provided very reliable, accurate service with rapid response time. They are more flexible than metal-sheathed thermocouples but also much for fragile. Non-sheathed thermocouples can not be used since the wiring itself would act as an antennae and transmit microwaves outside of the heating chamber. In general, the design of the heating chamber, thermometry and any metal internals should be carefully reviewed with a skilled microwave designer.

Microwave proponents often claim uniform heating but in reality non-uniform heating is a significant problem when heating materials that are strongly microwave absorptive. Microwave heating is volumetric heating (in theory it heats the entire volume of the object rather than conducting heat inward from the object's surfaces) but as the microwaves penetrate the surface their energy is absorbed proportional to the microwave field density and the absorptivity of the material. Penetration depth is greater at lower microwave frequencies.[4] When heating solid objects that contain large amounts of polar solvents, much of the heating will take place in the outer portions of the object as the polar material absorbs most or all of the microwaves. Electromagnetic CFD modeling is vital prior to fabrication and extensive experimentation. No practical industrial device exists for in situ measurement of microwave field density.[5] Some non-uniform heating can be mitigated by moving the solids (this is much less of a problem when heating liquids). However, when heating static, large solid objects, microwave non-uniformity is best reduced through electromagnetic CFD modeling, careful launcher design and testing at a moderate scale. A useful means of measuring the degree of non-uniform heating is thermal imaging. The figure below shows an example.

Figure 4 Thermal Image of Stack of solid objects heated with a 75kW Generator

When the objects being heating contain flammable materials, precautions must be taken to prevent a fire. Removal of two legs of the fire triangle is impossible since fuel is present and the potential for arcing is often present as long as microwaves as used (due to potential foreign metal objects or very high field densities). The key then, is to operate at reduced oxygen levels to prevent the potential of a fire. Consideration of leak potential and failure of the ceramic barrier window is important. Operation under vacuum could lead to air penetration so consideration for oxygen monitoring should be given. Failure of the barrier window during operation at positive pressure could lead to leakage of flammable chemicals into the waveguide providing a path to the high voltage magnetron. Inerting and the potential use of a second barrier window should be considered. In the event that double barrier windows are used, the captive volume between the barrier windows should be inerted at moderate pressure and this pressure monitored or the inert flowrate monitored. The figure below outlines a means of monitoring the integrity of the barrier window.

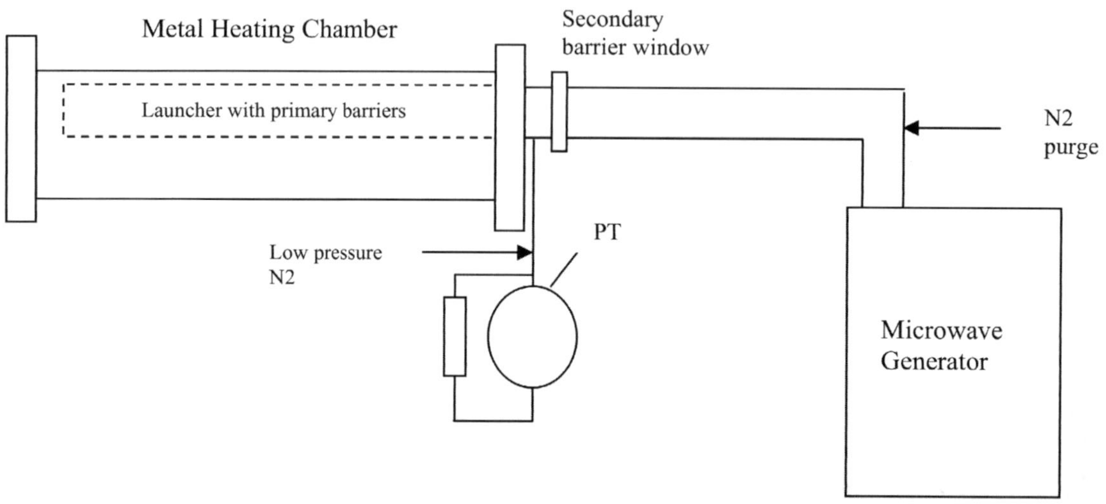

In summary, microwave heating can offer an efficient means of heating objects but an analysis should be carried out to determine if it is the lowest cost heating method. Microwave heating involves far more than merely connecting a microwave generator to a vessel. CFD modeling should be used to evaluate the potential outcomes and should influence the design for introducing microwaves into the heating chamber. The concepts of microwaves all follow principles of physics but they involve properties most chemical engineers have forgotten. Early in my experience with microwave heating I did not know enough about microwave heating to tell you what I didn't know. Finally, make sure you involve a competent high-power microwave designer.

My acknowledgements to:
Hal Kimrey Microwave Materials Technology
Leslie Montgomery Eastman Chemical Company
Chris Spencer Eastman Chemical Company

References:

1. Senise, Jose T. and Jermolovicius, Luiz A.: Microwave Chemistry – A Fertile Field for Scientific Research and Industrial Applications, Proceedings SBMO/IEEE MTT-S IMOC 2003.

2. Schiffmann, Robert F.: A Primer for Evaluating the Economics of a Microwave or Radio Frequency Processing System, Fourth World Congress on Microwave and Radio Frequency Applications, Austin, TX Nov. 2004.

3. Gerling, J.: Health, Regulatory and Equipment Safety Issues Related to Industrial Microwave Applications, Fourth World Congress on Microwave and Radio Frequency Applications, Austin, TX Nov. 2004.

4. Microwave Processing of Materials, Publication NMAB-473, 1994.

5. Kimrey, Hal, Microwave Materials Technology, Knoxville, TN, Private Communication.

Design and Operation of a Pilot Plant for Studies on Scale Up, Liquid Distribution, Liquid Holdup, Hydraulics, Heat and Mass Transfer, and Carbon Dioxide Removal Using Chemical Solvents

***Damaraju P. Rao**, Kaushik Majumder, and Srikanth Kandkuri*
Chemical Engineering Department, Indian Institute of Technology (Delhi),
Hauz Khas, New Delhi - 110016, India

Packed columns are widely used for gas-liquid contact operations in chemical industries as well as in environmental applications to accomplish separation of components in a given feed mixture. Knowledge of gas liquid parameters is important, specially in view of the advent of a large variety of new generation random and structured packings. The present work is aimed at collecting basic information on various transport parameters required for design or simulation of the absorption processes such as absorption and de-sorption of CO_2 into and from chemical solvents.

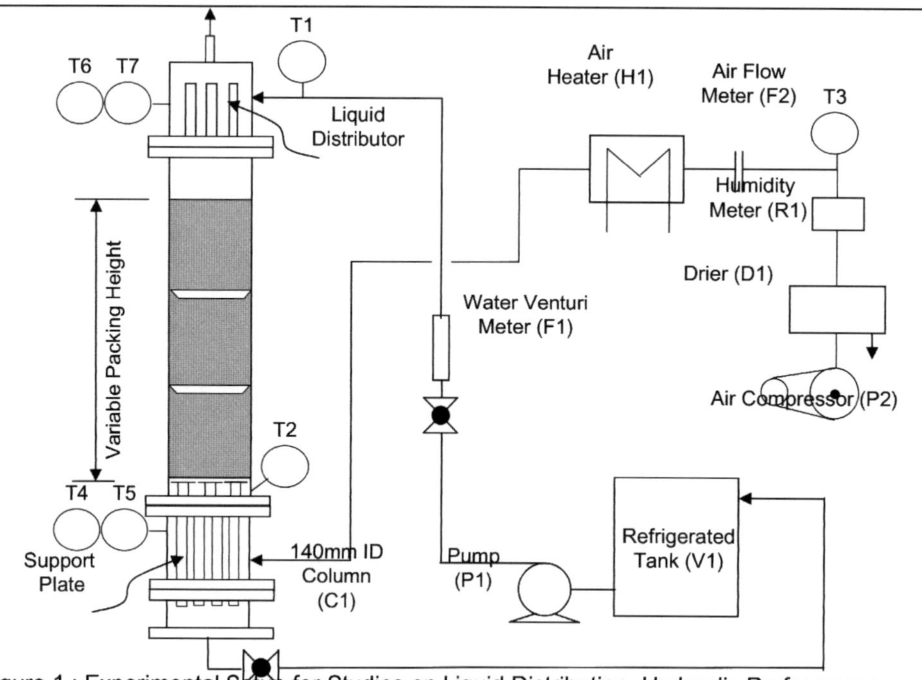

Figure 1 : Experimental Setup for Studies on Liquid Distribution, Hydraulic Performance, and Heat and Mass Transfer with Hot Air and Chilled Water

The following transport phenomena were studied experimentally with IMTP-15 packing in 140mm and160 mm diameter columns:

1. Liquid distribution at different liquid rates for different packing heights.
2. Liquid holdup, dry and wet pressure drop, and hydraulic capacity at different gas and liquid rates for different packing heights.

3. Gas phase heat and mass transfer between hot dry air and chilled water to realize nearly pure physical heat transfer conditions at different gas and liquid rates for different packing heights.
4. Gas phase heat and mass transfer between hot air-steam mixture and water at near boiling conditions to realize near pure mass transfer conditions at different gas and liquid rates.
5. Liquid phase residence time distribution as a function of gas and liquid flow rates
6. Liquid phase volumetric mass transfer coefficient as a function of gas and liquid flow rates under absorption and de-sorption conditions.

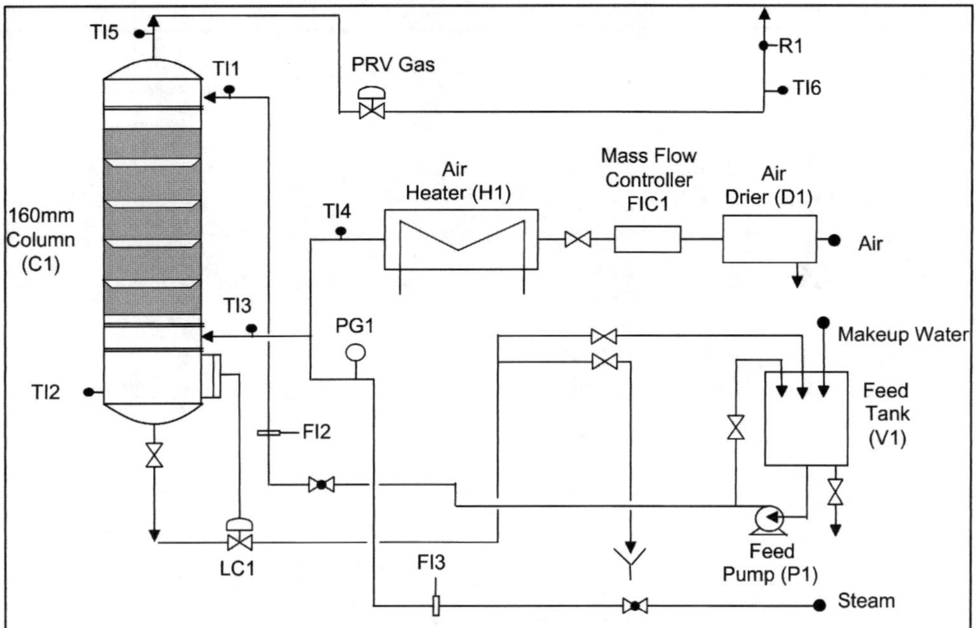

Figure 2 : Experimental Setup for Studies on Heat and Mass Transfer with Air-steam Mixture and Hot Water

Studies on liquid distribution, hydraulic performance, and heat and mass transfer with hot air and chilled water were carried out in a 140mm diameter Perspex column (Figure 1). Perspex column was selected to enable visual observation of liquid flow. The studies on heat and mass transfer with air-steam mixture and hot water were carried out in a 160mm diameter Stainless Steel column (Figure 2). All instruments used in different experimental setups were preciously calibrated against standard instruments to improve accuracy of experimental data and repeated experimental runs established reliability and repeatability of the measurement procedure.

The experimental setup for the study of liquid distribution consisted of column, water tank, water pump and water venturimeter. The liquid distribution was studied only with water because the gas flow was found to have negligible effect on liquid distribution. The column was made in multiple sections, so that the packing height could be varied. The column was equipped with liquid distributor, support cum collector plate and Rossette type wall wiper rings. The liquid distributor was riser-orifice type with 19 holes of 3.5mm diameter and 6 rectangular gas risers contributing to 20% of column cross-sectional area. The support cum collector plate was divided into three equal area annular zones to measure zonal distribution of liquid. With IMTP-15 packing in 140mm diameter column, the column to packing diameter ratio, that could be maintained was 9.3:1, which was on the lower side of the value of 10:1 to 12:1 as recommended by Mullin (1). Therefore, higher degree of wall flow was expected. Wall wiper rings were installed to correct it to the extent possible. Experiments were carried out for

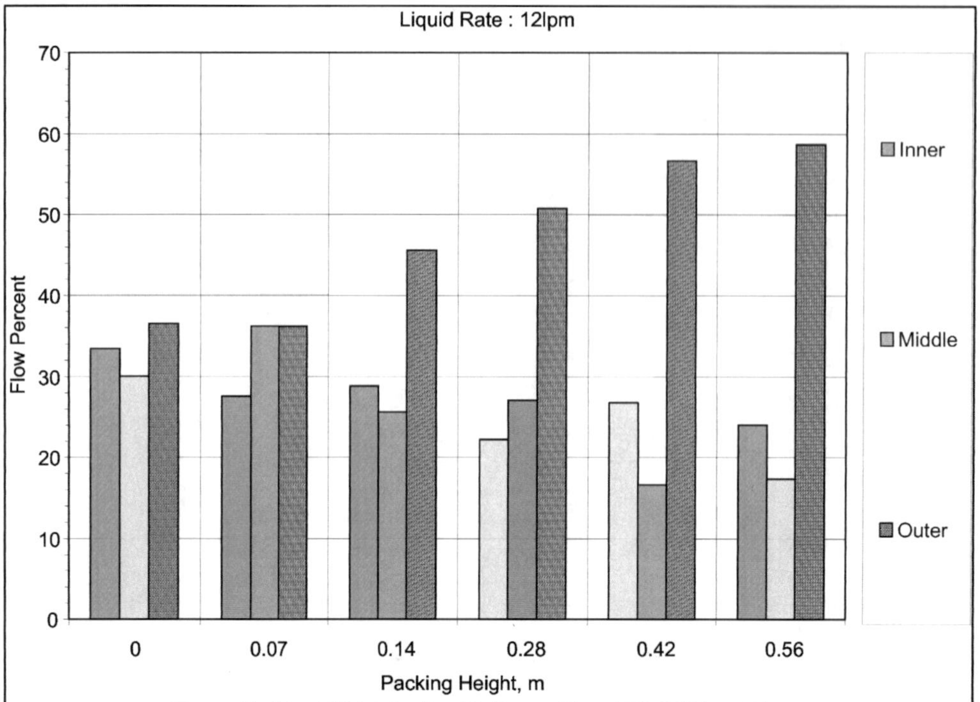

Figure 3 : Liquid Distribution Pattern without Wall Wiper Ring

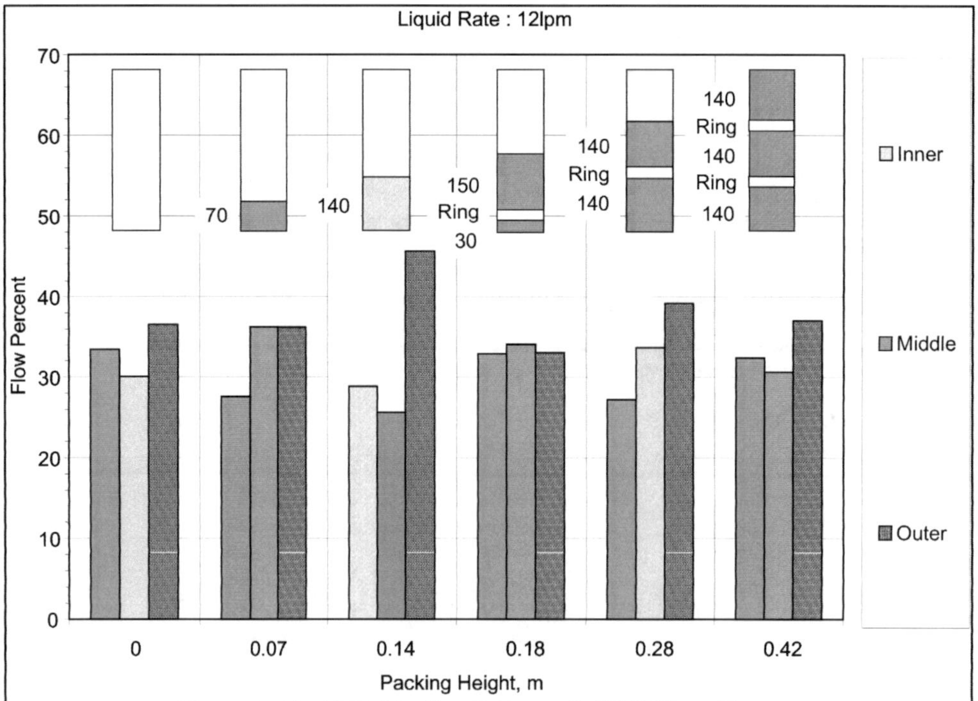

Figure 4 : Liquid Distribution Pattern with Wall Wiper Rings

different packing heights form 70mm to 560mm, first without using any wall wiper ring and then installing a number of wall wiper rings along the packed bed. From the first set of experiments, it was observed that without any wall wiper ring, wall flow was significant, and the wall flow increased with axial distance from the water inlet of packed bed to a distance from the water inlet at which the flow pattern stabilized (Figure 3). The next set of experiments showed that the wall flow was greatly reduced by using wall wiper rings and nearly uniform liquid distribution could be maintained by installing wall wiper rings at every height equivalent to one column diameter along the packed bed (Figure 4).

The experimental setup for liquid distribution study was modified for hydraulic studies by adding air compressor, air orifice meter, water and air temperature indicators, and water manometer to it. The design of the support plate was changed, as annular collection chambers were no more required. The support plate was redesigned to provide mechanical support to the packing, uniform gas distribution across the column cross-section, and separate flow paths for gas and liquid streams.

Hydraulic studies were carried out with four packing heights: 280mm, 420mm, 560mm and 700mm. At every 140mm of packing height, one wall wiper ring was placed. Water flow rates were maintained at 8lpm, 10lpm, and 12lpm. Air flow rate was varied between 300–1000lpm. For each packing height, dry pressure drop was first measured. Wet pressure drop was then measured at three liquid rates. At each liquid rate, airflow was increased till the flooding point was reached. The flooding point was defined as the point where liquid started appearing on the top of the packing followed by a sharp increase in liquid holdup and pressure drop and subsequent unstable functioning of the column. Experiments were also carried out for estimating the liquid holdup for four packing heights. For measuring liquid holdup, column operation was first stabilized at predetermined gas and liquid flow rates. Then, at one instant,

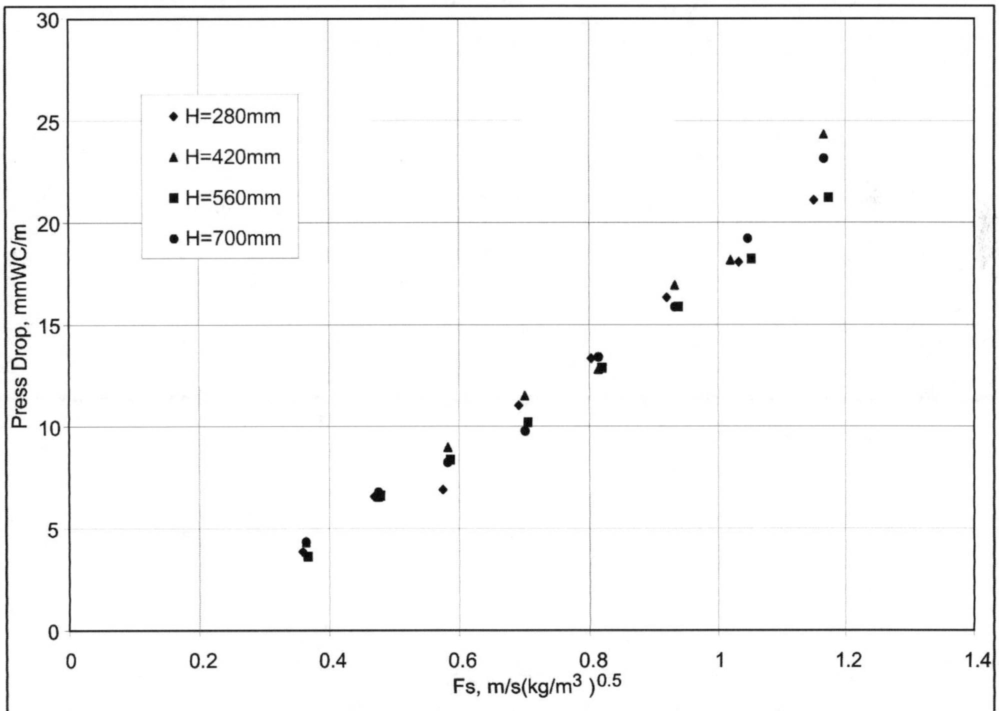

Figure 5 : Dry Pressure Drop for Different Packed Bed Heights (H)

liquid flows to and from the column and gas flow to the column were cut off and accumulated liquid in the column was drained and measured. Liquid holdup was also measured for no gas flow condition, which essentially estimated pre-loading liquid holdup because it was observed that below loading point, effect of gas flow on liquid holdup was negligible. The following observations could be made from the experimental data:

- Dry pressure drop (Figure 5) and pre-loading liquid holdup (Figure 6) were not affected by height of packed bed.

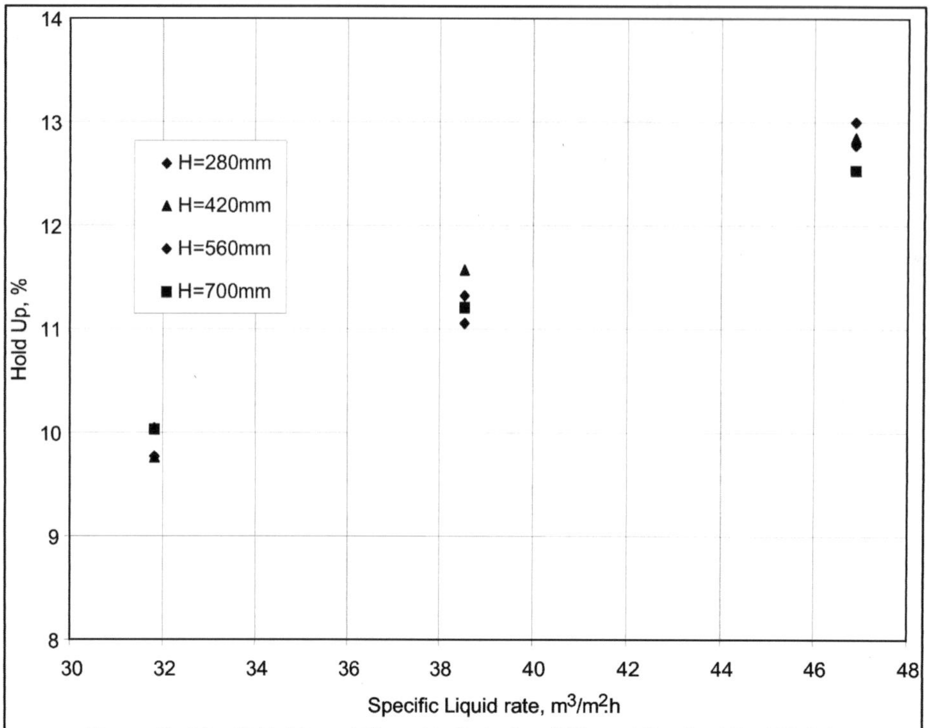

Figure 6 : Liquid Holdup at Zero Air Rate for Different Packed Bed Heights

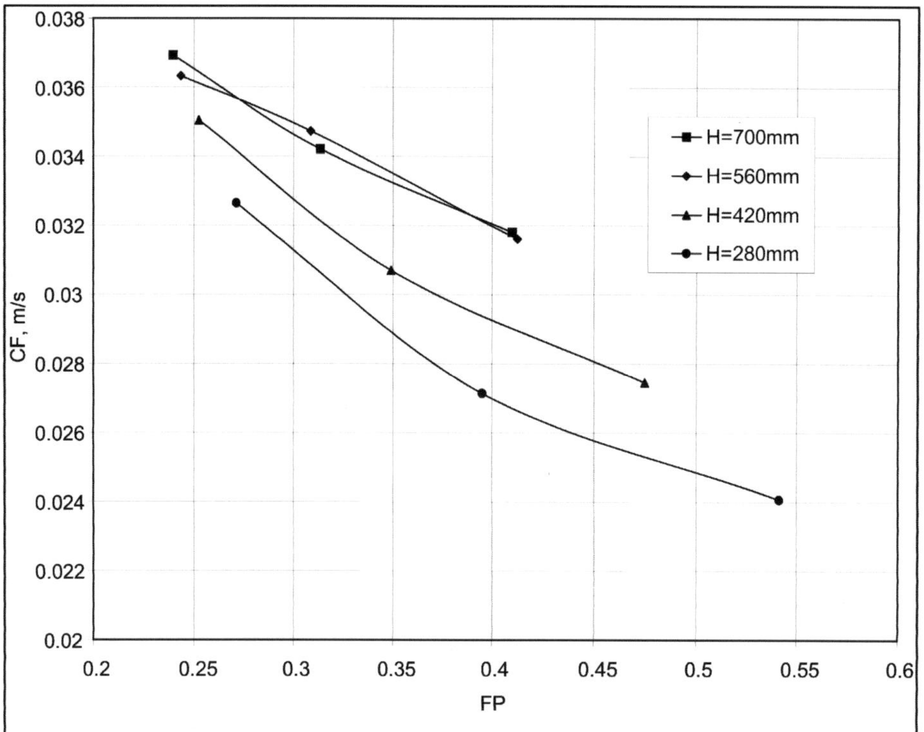

Figure 7 : Hydraulic Capacity at Different Packed Bed Heights (H)

- Flooding capacity (Figure 7), wet pressure drop (Figure 8), and liquid holdup beyond loading point (Figure 9) were affected by height of the packed bed. The effect was found to be significant in the case of shallow beds but gradually diminished with increasing bed height.

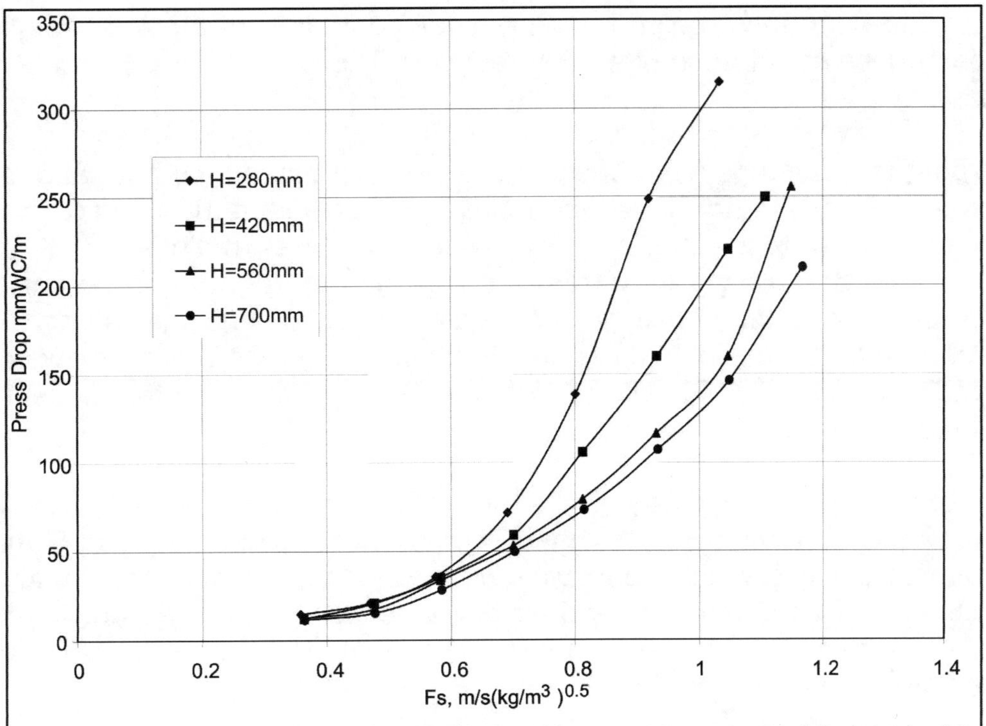

Figure 8 : Wet Pressure Drop at 32 m^3/m^2h Liquid Rate for Different Packed Bed Heights(H)

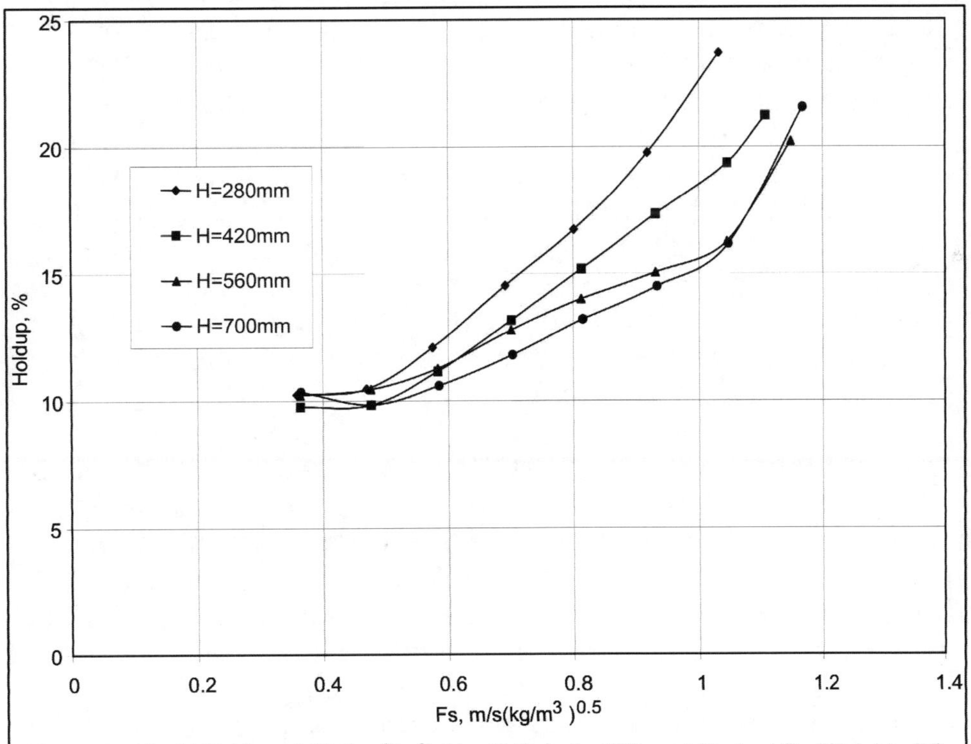

Figure 9 : Liquid Holdup at 32.0 m^3/m^2h Liquid Rate for Different Packed Bed Heights (H)

Models for pressure drop, liquid holdup, and capacity incorporating the effect of packing height were developed from the experimental data. The effect of packing height on hydraulic performance was attributed to gas phase entrance effect, which was due to strong interaction between gas and liquid at the gas entrance i.e. at the bottom section of packed bed. The effect was significant in shallow beds but with increasing bed height, as the gas flow got stabilized and the length of the entrance region became small compared to the total length of the bed, the relative contribution of entrance effect got reduced. The detailed analysis was presented by Majumder and Rao (2).

Two experimental setups were designed to study simultaneous heat and mass transfer. To conduct experiments with hot dry air and chilled water, the setup used for conducting the hydraulic studies was further modified. The water tank was maintained at near freezing temperature by circulating a refrigerant from a refrigerator through a cooling coil, compressed air was dried in a drier and then heated in an electric heater before passing it into the column. RTD type temperature indicators were used to measure temperatures of water and air at column inlet and exit. Humidity of inlet and exit air was estimated from the measurement of dry bulb and wet bulb temperatures.

Experiments with hot dry air and chilled water were carried out with six packing heights: 70mm, 140mm, 280mm, 420mm, 560mm and 700mm. For packing heights 280mm and more, at every 140mm of packing height, one wall wiper ring was placed. Water flow rates were maintained at 8lpm, 10lpm and 12lpm and air flow rate was maintained between 300lpm and 1000lpm. Inlet water temperature was maintained between 1°C and 2°C. Inlet air temperature was maintained between 65°C and 75°C.

For heat and mass transfer studies with air-steam mixture and hot water, a new experimental setup was created. The setup consisted of 160mm diameter S.S. column, water tank, water pump, air compressor, air drying system, air heater, and steam boiler. Water flow was measured by electromagnetic flow meter, air flow was measured and controlled by air mass flow controller, and steam flow was measured by vortex flow meter. RTD type temperature indicators were used to measure temperatures of various process streams in the setup. The column was fitted with pressure gauge at gas inlet and pressure reducing valve at gas exit. Level controller was installed at the column bottom sump. Humidity of gas at column exit was measured by humidity meter. Column inlet water temperature was maintained constant by circulating the water in a closed loop through a circulation tank in which cold water in required amounts was added and excess water drained to maintain the tank level between limits.

Experiments with air-steam mixture and hot water were carried out with 1m packing height and at every 160mm of packing height, one wall wiper ring was placed. Water flow rates were maintained at 10lpm, 13lpm and 16lpm, which were equivalent to same specific liquid loads, maintained in 140mm diameter column. Air flow rate was maintained between 300Nlpm and 800Nlpm. Steam rate was maintained between 150lpm and 615lpm. Experiments were conducted at atmospheric pressure. A few runs were taken at elevated column inlet pressure of about 30psig. Inlet water temperature was maintained between 77°C and 85°C. Preheated air temperature was maintained close to saturation temperature of steam so that the air-steam mixture temperature at the column inlet could be maintained between 110°C and 122°C.

A mathematical model for simultaneous heat and mass transfer based on film theory and assumption of plug flow of gas and liquid that predicts the gas and liquid temperature profiles as well as humidity profiles in the column at steady state was used to estimate gas phase heat and mass transfer coefficients from experimental measurements. The estimation method consisted of minimizing the sum of squares of differences between the measured and predicted outlet conditions of the gas and liquid streams. From the analysis of data with the help of the model, it was observed that the coefficients were dependent on gas flow rate and height of packed bed. The effect of packing height on the values of the coefficients was significant when the height was low and it was attributed to gas phase entrance effect (Figure 10). Similar influence of packing height was observed on hydraulic parameters. The model parameters derived for incorporating the effect of packing height on hydraulic, heat and mass transfer coefficients showed similar trend with respect to variation in packing height.

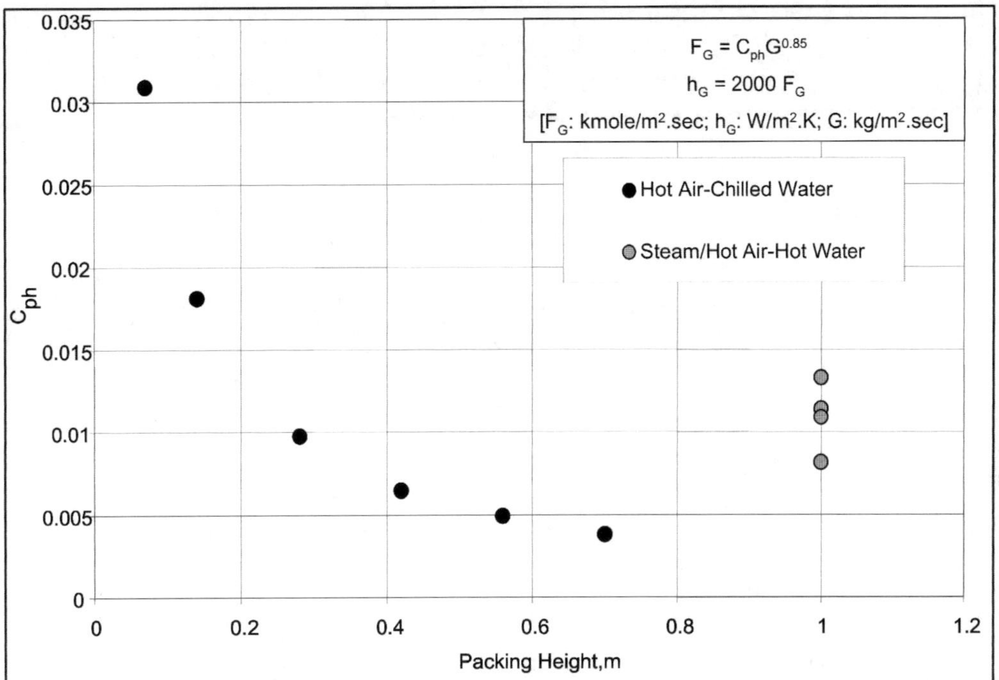

Figure 10 : Variation of Gas Side Heat & Mass Transfer Coefficients with Packing Height

Step change response experiments on a 140mm diameter, 560mm tall column packed with IMTP-15 were conducted to determine the liquid phase residence time distribution for different

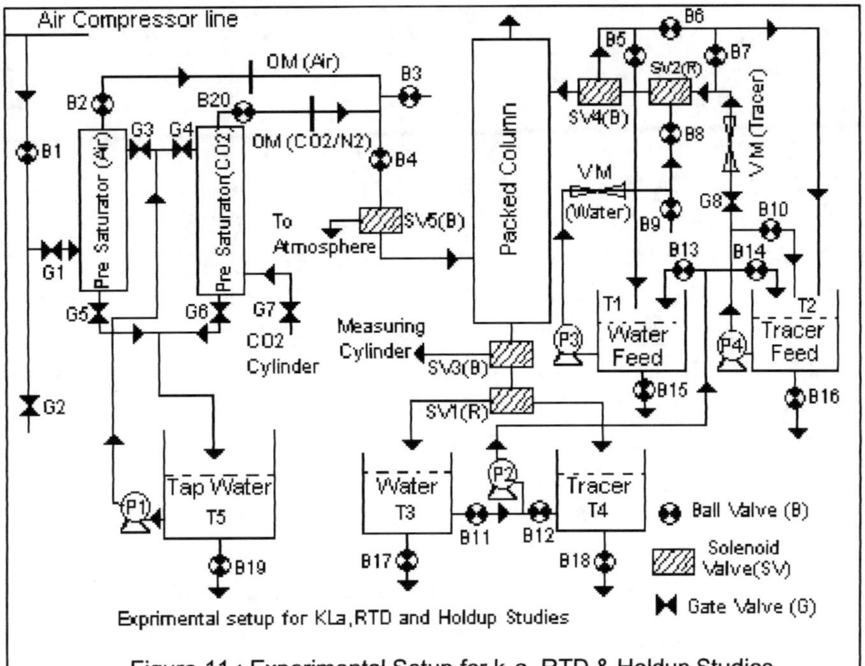

Figure 11 : Experimental Setup for k_La, RTD & Holdup Studies

values of gas and liquid flow rates (Figure 11). The design of the column was the same as that used for studying liquid holdup, and gas phase heat and mass transfer. Solenoid valves were fitted in the apparatus shown in Fig. 11 to enable determination of dynamic liquid holdup, and give step change in the feed concentration of tracer for RTD studies. Feed gas was presaturated to prevent evaporation of water in the column. The extent of static hold up was assumed to be negligible and axial dispersed plug flow model was used to determine Bodenstein number from step response data. The estimated values of Bodenstein numbers were found to have increased in the range 20-30 with increase in liquid flow rate and no effect of gas flow rate was found till a value of air flow rate equal to 800Nlpm (Figure 12). After inserting the liquid redistributors in the packed column, the Bodenstein numbers were found to have increased marginally by about 10%.

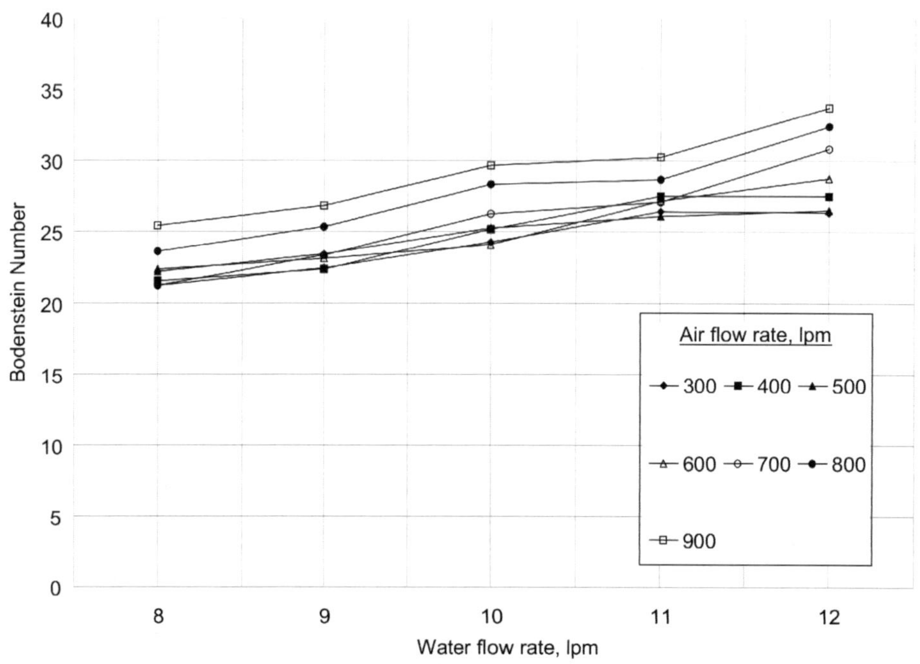

Figure 12 : Bodenstein Number vs Water Flow Rate at Various Air Flow Rates

The liquid phase volumetric mass transfer coefficient was estimated by conducting experiments in which pure carbon dioxide gas was absorbed into water in a packed column of diameter 140mm and height 560mm. The extent of absorption was indirectly monitored by measuring the raise in electrical conductivity of water that occurs due to formation of carbonic acid. A mathematical model for steady state conditions that takes axial dispersion in the liquid phase into account was used to determine the volumetric mass transfer coefficient by matching the predicted raise in electrical conductivity with the corresponding experimental value. The liquid phase mass transfer coefficient data were found to increase with increase in liquid flow rate and no effect of gas flow rate was found (Figure 13). The insertion of Rosette type of liquid redistributors was found to have caused an increase in the value of the volumetric mass transfer coefficient by about 20%. The values of volumetric de-sorption coefficient were also estimated by desorbing carbonic acid solution into air in the packed column and measuring the drop in electrical conductivity of water at steady state from inlet to outlet. A mathematical model based on the assumption of axial dispersion in the gas and liquid phases was used to estimate the de-sorption mass transfer coefficient. The value of axial dispersion coefficient for the gas phase was estimated from literature data. The values of de-sorption mass transfer coefficients were found to be much higher than the values of absorption

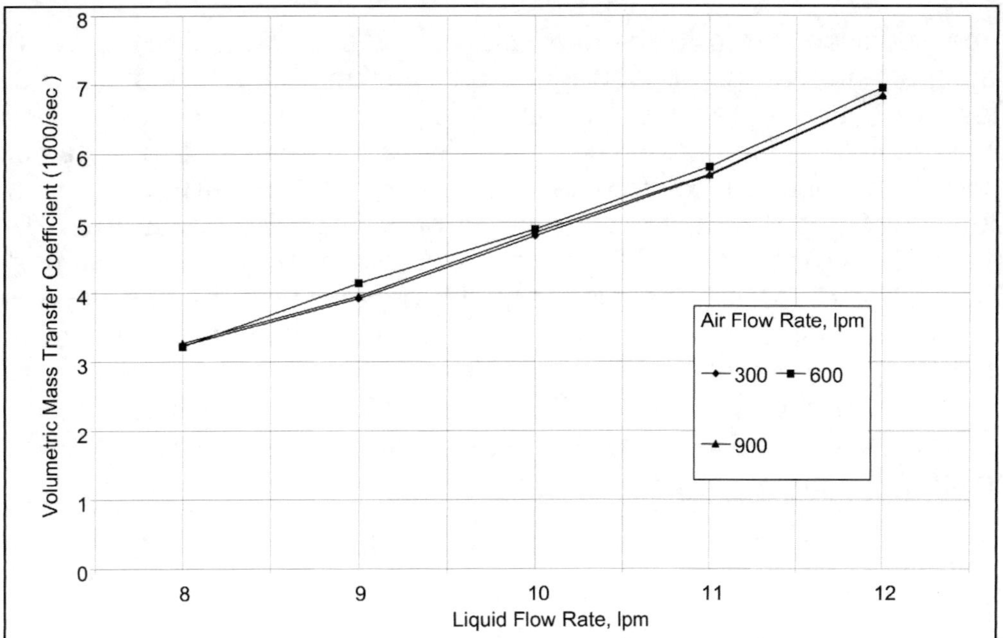

Figure 13 : k_La (absorption) vs. Liquid Flow Rate at Various Gas Flow Rates

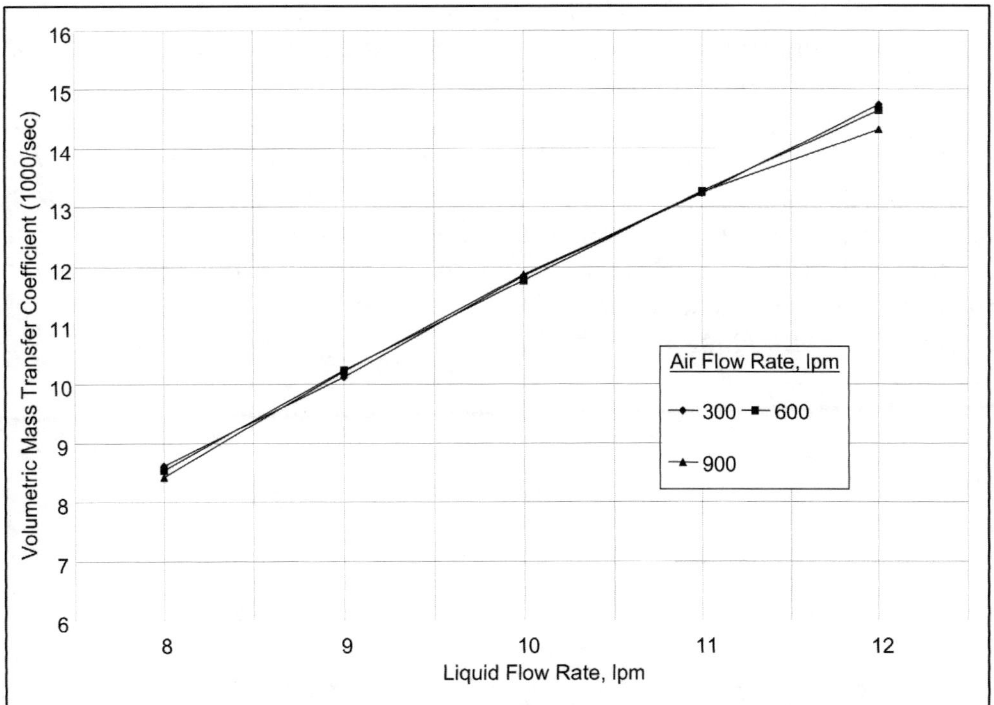

Figure 14 : k_la (de-sorption) vs. Liquid Flow Rate at Various Gas Flow Rates

mass transfer coefficients for identical values of gas and liquid rates (Figure 14). The reason for the above observation was given on the basis of Brian's theory of super-saturation that results when the diffusivity of the absorbing gas (oxygen in the present case) is higher than that of the desorbing carbon dioxide gas. The above observation on de-sorption process was made earlier by Panja and Rao (3,4) in their studies on mechanically agitated contactors and Thaker and Rao (5) in their studies on bubble columns.

It can be concluded from the studies made in the present work that redistribution of liquid, and gas sparging through support plate, and packing height for shallow columns have important effects on various gas-liquid parameters. Such effects need to be confirmed by conducting experiments with larger dia. columns. Frequent redistribution of gas and liquid can have an important bearing on column heights required for gas-film controlled operations. Studies on 300 mm dia. columns are currently in progress. Gas phase axial dispersion and interfacial area with 140 mm dia IMTP and Sulzer Mellapak packed columns are also in progress. The pilot plant facilities developed in the present work can be used for evaluating packings and chemical solvents. In the present setup, complete automation of the pilot plant has been achieved with the help of an online computer interfaced to the process to facilitate precise setting of various operating conditions. The pilot plant facilitates thus validation of mathematical models, estimation of model parameters, study of scale-up rules, and even advanced control algorithms.

References:

1. Mullin, J.W., 1957, The Effect of maldistribution on the performance of packed columns, *The Industrial Chemist*, 33: 408.
2. Majumder, K. and Rao, D.P., 2005, Effect of Bed Height on Liquid Distribution and Hydraulic Performance of Packing, *58th Annual Session of Indian Chemical Engineering Congress, Delhi, December, 2005.*
3. Panja, N. C. and Rao, D.P., 1993, Measurement of gas-liquid parameters in a mechanically agitated contactor, *The Chem. Engineer*, 52: 121.
4. Panja, N. C. and Rao, D.P., 1994,Volumetric mass transfer coefficients from de-sorption data, *Chemical Engineering Research and Design*, 72: 633.
5. Thaker, K and Rao, D.P., Effects of gas-redispersion and liquid height on gas-liquid hydrodynamics in a multistage column, a paper under consideration for publication in *Chemical Engineering Research and Design*.

Analysis of Distillation Differential Equations for Improved Operational Policies and Equipment Configuration

Lubilanji Jean Mulopo[1], Diane Hildebrandt[2], and David Glasser[2].

(1) School of Chemical Engineering, Howard College, University of Kwazulu-Natal, Durban 4041, South Africa,
(2) Chemical and Metallurgical Engineering, University of the Witwatersrand, 1 Jan Smuts Avenue, Braamfontein, Johannesburg, South Africa

Doherty [1] introduced the concept of differential equations (DE's) as a shortcut design tool to determine the composition profiles along the length of the rectifying and stripping sections in a distillation column. Each section has a differential equation that describes the change of liquid composition along the column section. Equimolar overflow is assumed for simplicity, although the analysis can be extended to include non constant molar overflow. The advantage of this model is that it allows finite reflux ratios (reboil ratio) to be analysed. A differential equation model for the stripping section of the column is:

$$\frac{dx}{ds} = (y - x) + \Delta(x_b - x) \qquad (1)$$

where s is the integration variable and x_b, the bottoms composition. The differential equation in this case is solved so that the profile starts at the bottoms composition (i.e. s=0 at $x=x_b$) and continues upwards in the column towards a feed composition until it reaches its pinch point. One notices that the integration variable s is a function of the distillation energy input and number of stages.

For a binary mixture with constant relativity, if one keeps the reboil ratio constant this will lead to an increasing energy demand to reach the same bottom composition. Alternatively the reboil ratio can be manipulated during the process to balance the change in energy requirements. In this regard, a number of alternatives are usually considered. Thus there is need of simple and quick assessment tools to compare these alternative options. In this paper we show that the analysis of distillation differential equations can provide a basis for the ranking of these alternative processes. Using a case study we show how the integration variables in these differential equations can be used as optimization parameters for improved operational policies in terms of energy demand and/or choice of equipment configuration. We believe this is a new way of looking at process synthesis of distillation systems.

References

[1] Doherty, M.F. and Malone, J.D.," Conceptual Design of Distillation Systems" McGraw Hill, New York, 2001

[2] Holland, S.T., Tapp, M., Glasser, D., Hildebrandt, D.," Novel separation system design using "moving triangles" Computers chem. Eng., 2004, 29, pp.181-189

Visualisation of Mass Transfer across an Electrically Charged Plane Interface

Zheyan Qiu, CPE, The University of Kansas, Learned Hall, 1530 W 15th Street, Lawrence, KS 66045, Jerzy Petera, Process and Environmental Engineering, The Technical University of Lodz, ul Wolczanska - 175, 90-929 Lodz, Poland, and Laurence R. Weatherley, Chemical and Petroleum Engineering, The University of Kansas, Learned Hall, 1530 W. 15th St. Room 4132, Lawrence, KS 60045-7609.

The intensification of liquid-liquid transport processes using externally applied electric fields is well known. The presence of electric charges or local variations on their surface density will result in local variations in the effective interfacial tension and thus lead to interfacial tension induced surface turbulence (Maragoni instability). Such phenomena increases mass transfer rate. Promotion of interfacial disturbance is especially interesting because a number of novel liquid-liquid systems such as those involving homogeneous catalysts, or involving enzymes which are inter-facially active potentially could be intensified by the application of electrical fields. This paper is concerned with the fundamental interfacial mechanisms behind the observed increase in mass transfer across a plane interface in the presence of an electric field. Experimental study of this for some real liquid-liquid systems is presented, and a novel technique for measuring the rate of interfacial mass transfer is described. Schlieren technique was utilized to record visually the hydrodynamic disturbances in the vicinity of the liquid-liquid interface. The following experimental variables are included in the study: The electrical charge on the plane interface; the magnitude and direction of the external electrical field; viscosity and interfacial tension of the system; the presence of surface active agents. Simultaneously, finite element modelling of interfacial mass transfer was used to predict significant time-dependent convective disturbances at the liquid-liquid interface.

Process Intensified Green Reaction Chemistries Performed in the Spinning Tube-in-Tube (Stt®) Reactor

Michael A. Gonzalez, US Environmental Protection Agency, 26 W Martin Luther King Drive MS 466, Cincinnati, OH 45268 and Phillip L. Lichtenberger, Kreido Laboratories, 1140 Avenida Acaso, Camarillo, CA 93012.

The U. S. Environmental Protection Agency (USEPA) and Kreido Laboratories have established a Cooperative Research and Development Agreement (CRADA) collaboration, to develop and commercialize green and sustainable chemistries in the area of industrial chemical synthesis. Utilizing the patented and proprietary STT® reactor technology designed and developed by Kreido Laboratories, the STT® reactor utilizes process intensification techniques to increase the rate, yield, and selectivity of chemical reactions used in producing commercially important materials. The USEPA has broad experience and expertise in designing and developing environmentally friendly chemical processes which can lead to the implementation of green and sustainable reaction chemistries. This collaborative effort has lead to the development of reaction chemistries in the areas of oxidation, hydrogenations and esterifications. Current research results and the benefits observed by these new green and sustainable chemistries in this novel reactor will be presented.

Prediction of Pressure Drop and Liquid Holdup in Katapak-Sp Using a Particle Model

Ricardo Macías-Salinas, ESIQIE-Chemical Engineering Department, Instituto Politécnico Nacional, UPALM, Av. IPN s/n, Col. Lindavista, México, D.F., Mexico, Laura Rosas-Ortiz, ESIQIE-Petroleum Engineering Department, Instituto Politécnico Nacional, UPALM, Av. IPN s/n, Col. Lindavista, Mexico, D.F., Mexico, and Elisabetta Brunazzi, Chemical Engineering Department, University of Pisa, Via Diotisalvi 2, Pisa, Italy.

Various experimental efforts have been reported in the literature in an attempt to understand the hydrodynamic behavior of a relatively new family of catalytic structured packings: KATAPAK-SP. The only approach so far used to represent the experimental results obtained for KATAPAK-SP 11 & 12 relies in the channel model. The purpose of this work was to verify the suitability of a particle model originally developed for random and structured packings (Stichlmair et al., Gas Sep. Purif., 3, 19, 1989) by extending its applicability to catalytic structured packings, particularly for the KATAPAK-SP family. In doing that, a more suitable liquid holdup correlation for the open channels was proposed. The question on how liquid splits into the open channels and catalyst bags still remains partially unanswered. In an attempt to solve this, a new calculation procedure has been devised here to reasonably determine the liquid share between the open channels and the catalyst bags. Model predictions in terms of pressure drop and liquid holdup are presented and discussed in this work for the case of KATAPAK-SP 11 & 12.

Studies on ionic mass and momentum transfer with coaxially placed twisted tape - disc assembly as turbulence promoter in circular conduit

Nageswara Rao, V[1]., Rajendra Prasad, P[2], Sujatha,V[3].,Vaka Murali Mohan[4]
Department of Chemical Engineering, College of Engineering,
Andhra University, Visakhapatnam-530 003, Andhra Pradesh, INDIA.

Abstract:

Studies on the effect of coaxially placed entry region twisted tape-disc assembly as turbulence promoter on mass transfer rates in forced convectionflow of electrolyte were conducted. The study comprises of evaluation of mass and momentum transfer rates at the outer wall of the electrochemical cell. Pressure drop measurements were also made to obtain the energy consumption pattern. Mass transfer coefficients were evaluated from the measured limiting currents. The study covers a wide range of geometric parameters such as pitch of the tape (T_P), length of the tape ((T_L) , width of the tape (T_W), diameter of the disc (D_d), tape-disc distance (h). The results revealed that the mass transfer coefficient increases with increase in velocity, diameter of the disc (D_d), length of the tape (T_L), width of the tape (T_W) and decreases with increase in pitch of the tape (T_P) and tape-disc distance(h). Within the range of variables covered, the augmentations achieved in mass transfer coefficients were up to 5 fold over the tube flow in absence of promoter. The data on momentum transfer were analyzed with momentum transfer roughness function and roughness Reynolds number. Mass transfer rates were analyzed with mass transfer roughness function and roughness Reynolds number. The following correlations were reported out of the study.

$R(h^+)=3.573\ (Re^+)^{0.068}\ (\phi_1)^{0.050}\ (\phi_2)^{-0.033}\ (\phi_3)^{0.012}\ (\phi_4)^{0.228}\ (\phi_5)^{-0.037}$

$g(h^+) = 343.95\ (Re^+)^{-0.176}\ (\phi_1)^{0.244}\ (\phi_2)^{-0.001}\ (\phi_3)^{-0.033}\ (\phi_4)^{0.402}\ (\phi_1)^{0.016}\ Sc^{-0.345}$

INTRODUCTION

Process intensification is a never-ending desire for process engineers. It has been carried out either for efficient operation or to bring down production cost to a minimum. To that effect, process engineers are improving their process all the time. Electrochemical reactor design is not an exception to this norm. To achieve this task a number of approaches had been resorted to. Use of turbulence promoter is one method resorted in general. The turbulence promoters are of several kinds. Application of additives to the reaction system, adaptation of mixing devices is principal technique. Among them use of baffles(1), stirrers(2), liquid jets(3), spargers(4), and various other geometrical bodies immersed in the flow and the use of two phase(5), three phase flows were commonly adopted techniques. The list was long and the visualization of a technique is the creativity of a researcher. Uses of geometrical objects in the flow streams have been receiving the attention of the several workers. Bergles (6) reviewed several techniques and types of promoters. The augmentations reported to be several hundred fold and the associated pressure losses were also found to be not insignificant. However, these studies emphasized major advantages of reduction in equipment size, floor space, and higher throughputs. Ease of fabrication, assembling and installation also offer remarkable cut in the initial investment costs. Use of displaced promoters or insert promoters had been gaining importance. Augmentation in mass transfer coefficients were reported with the use of string of spheres(7), string of discs(8), spiral coils(9), coaxially placed cones(10), twisted tapes mounted on a central annular rod(11). Present study uses entry region twisted tape-disc assembly as insert promoters. Discs(12) and orifices(13) placed across the flow in conduit generate form friction both in upstream and downstream sides, thus generated turbulence continues to show their effect on wall mass transfer coefficients to longer lengths along the cell. The flow fields generated by this type of promoters could not be explained properly. The enhancement of mass transfer coefficient was satisfactory at a distance from the disc under consideration. The studies mentioned above enhanced mass transfer rates to a reasonable extent, on the other hand the transfer coefficients obtained were not uniform along the length of the reactor. Moreover, they showed oscillatory behavior with distance. The generated oscillations were of decaying nature along the path of the flow. Efforts are also directed towards the use of compound augmentation in order to achieve maximum possible throughputs at minimum power loss and pumping costs. Twisted tapes and discs are useful insert promoters. In earlier investigations where twisted tapes as entry region swirl generators and string of discs were employed as promoters, significant augmentation in heat and mass transfer were reported. Studies on the effect of coaxially placed twisted tape-disc assembly in circular conduits on mass transfer rates in case of forced convection flow of electrolyte have not been reported. Present investigation is therefore undertaken and essentially deals with the evaluation of mass transfer rate at the outer wall through limiting current technique. Parameters covered in the study are compiled in Table - 1

Table - 1.

Range of variables covered in the study.

Variable	Minimum	maximum	Max/Min
Pitch of the Twisted Tape, m/turn	0.02	0.12	6
Length of the Twisted Tape,m	0.14	0.22	1.57
Width of the Twisted tape,m	0.01	0.03	3
Diameter of the of Disc,m	0.02	0.045	2.25
Distance between Tape and Disc,m	0.05	0.25	5
Velocity,m/s	0.024	0.224	9.33 .
Reynolds number,Re	1285	11985	9.32
Schmidt number, Sc	830.8	1251	1.50

EXPERIMENTAL

The experimental setup used for the study was similar to those used in earlier studies(14,15,16). Data on limiting currents for the case of reduction of ferricyanide ion is obtained for fluid flow in circular conduits in the presence of twisted tape-disc assembly as insert promoter and was shown in Figure 1.The following electrode reaction was involved in the study.

Cathodic reduction of ferricyanide ion:

$[Fe(CN)_6]^{-3} + e$ ------------ $[Fe(CN)_6]^{-4}$ .... (1)

Equimolal solution of 0.01 M Potassium ferri- ferrocyanide couple was chosen along with 0.5N NaOH were prepared. The point electrodes fixed flush with surface of the wall of the cell were used to obtain limiting currents. Initially blank runs were conducted with sodium hydroxide solution alone to ensure that the limiting currents obtained in the subsequent runs are due to diffusion of reacting ions (Ferri cyanide ion) only. The electrolyte was pumped at a desired flow ratethrough the test section by operating the control and by-pass valves. On attainment of steady state, potentials were applied across the test electrode and wall electrode in small increments of potentials (100mV) and the corresponding currents were measured for each increment. In view of the large area of the wall electrode in relation to the test electrode nearly constant potential is maintained at the test electrode. The limiting currents were obtained from the measurements of applied potential and current as had been done in several earlier works(14,15,16). The attainment of limiting current was indicated by the constancy of current with a large increase in the potential. Mass transfer coefficients were computed from the measured limiting currents by the following equation:

$k_L = i_L/nFA_0C_0$ ....(2)

Pressure drop measurements were also taken simultaneously using an inclined manometer with Carbon tetrachloride as manometric liquid.

RESULTS AND DISCUSSION

It is evident from the earlier studies twisted tapes augment transfer rates(17). Transfer rates were also enhanced by the disc(12) placed across the flow inside the conduits. Energy consumed for these processes were of several folds. An attempt was made to find synergic effect of these two promoters. Transfer rates were found to vary along the test section and depends upon the region in which it falls because of localized nature of turbulence. The transfer rates were computed for each point electrode as they were placed all along the length of the test section. The test section was divided into two different regions depending upon the placement of the promoting elements. The two regions namely were 1) tape-disc region and 2) region down stream to disc. Tape disc region was an important part of the test section as the turbulence and hence transfer rates are prominent. The region down stream to disc is termed as down stream region. The turbulence developed in the tape-disc region was transported along the test section to the downstream region. Transfer rates normally decline rapidly in this region.

Average limiting currents were computed by taking arithmetic average of limiting currents obtained from point electrodes. The average limiting currents obtained by graphical integration technique was in close agreement with those calculated from arithmetic average. The effect of tape-disc promoter on mass transfer was clearly observed. The average limiting current values obtained were used for analysis of mass transfer data. Introduction of co-axially placed twisted tape in circular conduit alter the flow pattern in cell by imparting swirling action to the axial flow. The swirl in turn influences tractive shear forces near the wall region which in turn reduce the thickness of the concentration boundary layer there by augment mass transfer rates. From the literature (12), it was evident that the introduction of bluff bodies like co-axially placed circular discs augment mass transfer rates considerably. Therefore, the introduction of twisted tape in the entry region together with a co-axially placed disc away from the twisted tape was expected to augment mass transfer rates considerably. Individual effects of tape and disc were shown in the figure 2. The synergic effect of tape disc promoter was found better.

Effectiveness of promoter

Any turbulence promoter is expected to generate turbulence and thereby enhance mass transfer rates. The augmentation obtained is an important tool for the evaluation of effectiveness of promoter. Mass transfer coefficient was plotted against velocity and shown in figure 3. The plot A in the figure is the data of Lin et al for smooth pipe flow and with no internals. The plots B and C correspond to the sets that has combination of length of tape (T_L) pitch of tape (T_P) width of tape (T_W) diameter of disc (D_d) and distance between tape and disc (h) which yielded the minimum and maximum augmentation respectively. Observation from the figure 3 indicates that at lower velocity(0.024m/s) the augmentation found to be 2.94 times to 9.48 times over Lin et al. Similarly for a higher velocity, 0.224m/s the augmentation found to be 1.1 times to 4.4 times over Lin et al. for the least and the most augmenting promoter combinations respectively. From these observations it was found the synergic effect of tape-disc promoter was considerably high. Effect of individual parameters was discussed here under:

Effect of pitch of tape:

The pitch in all these experiments was defined as distance between two consecutive turns (m/turn) and this is varied from 0.02m/turn to 0.12m/turn. Figure 4 shows the variation of k_L vs V when tape- disc assembly was present in the conduit with pitch as parameter. Mass transfer coefficient was decreasing with increase in pitch of the tape. As the pitch of tape decreases from 0.12m/turn to 0.02m/turn, mass transfer coefficient increased from 125 percent to 221 percent over those values predicted from Lin et al for smooth conduit at a velocity of 0.224m/s. Therefore the use of disc in combination with twisted tape has a synergic effect on mass transfer.

Effect of the length of the tape

Figure 5 demonstrates the effect of the length of the tape for the tape-disc assembly. The lengths of tapes used in the study were varied from 0.22m to 0.14m. Mass transfer coefficient was increased with the increase in length of tape. The increase in mass transfer coefficient was 189 percent at the tape length 0.22m while that is 169 percent at a tape length of 0.14m over smooth tube values at a velocity of 0.224m/s. These observations were consistent with other combination of geometric parameters of the tape-disc assemblies.

Effect of the width of the tape

Mass transfer coefficient was plotted against velocity for different tape width values, keeping other geometric parameters constant as shown in Figure 6. Tape width had pronounced effect on mass transfer coefficient by imparting swirl motion to the fluid in the conduit. Tape width alters the local velocity of fluid near the wall thereby increasing the mass transfer rate. Plot A in figure 6 was the data of Lin et al. The figure revealed enhancements in mass transfer varied from 1.36 to 1.80 times over Lin et al as the tape width varied from 0.01m to 0.03m at a velocity 0.224m/s.

Effect of disc diameter

The plots of figure 7 were drawn for k_L vs V to show the effect of disc diameter keeping all other geometric parameters constant as shown in the inset of the figure. Plot A was that of Lin et al. Plot B,C,D,E, and F are the curves for disc diameters of D_d=0.02m,0.025m,0.03m,0.035m and0.045m respectively. The augmentation in mass transfer was 4.6 times for the disc diameter 0.045m at a velocity 0.224m/s while the augmentation was 3 times for the disc diameter 0.02m over smooth tube values. It was observed that disc diameter had drastic effect on the mass transfer process. As the disc diameter increases mass transfer increases.

Effect of distance between tape and disc

When a disc was placed across the flow immediately after the tape, mass transfer rates augmented at considerable levels. Plots in figure 8 were drawn for k_L against velocity for different values of tape-disc distance (h) varying from 0.05m to 0.25m keeping all other parameters constant at T_P=0.035m, T_L=0.14m, T_W=0.015m, D_d=0.025m. Mass transfer coefficients were decreased from 2.2 times to 1.25 times over Lin. et. al,. while increasing the tape-disc distance from 0.05m to 0.25m at a velocity of 0.224m/s. It revealed that mass transfer coefficients showing a decline as the distance between tape - disc was increasing.

DEVELOPMENT OF CORRELATIONS:

Correlation of momentum transfer data:
A disc inserted across the flow of fluid through a conduit generates eddies, causes intense turbulence in the vicinity of the disc and extends to the wall. The developed turbulence extended from upstream to the down stream. As a result of these flow fields, the flow is fully turbulent causing a turbulent core on the central region of tube and viscous flow very near to the wall of tube.
From the measured pressure drop data, friction factor f was calculated using the following equation, for each region mentioned earlier.

$$f = \Delta P . d . g_c / 2 . L . \rho . V^2 \quad \text{......(3)}$$

An attempt was made to correlate the friction factor data with Reynolds number without taking into account of any geometrical parameters. The following equation was resulted with a standard deviation as high as 39 percent.

$$f = 0.152 \times 10^8 \, Re^{-1.915} \quad \text{...... (4)}$$

By the incorporation of the geometric parameters of the study. the following equation was resulted.

$f=0.339x10^8\ Re^{-1.918}\ (\phi_1)^{-0.289}\ (\phi_2)^{0.232}\ (\phi_3)^{-0.129}\ (\phi_4)^{-0.095}\ (\phi_5)^{0.265}$ ....(5)

Average deviation =17.67, Standard deviation =37.32

Even this equation obtained for the f vs Re including geometrical groups showed large deviation. Therefore an alternative approach was resorted to fit the data with another model. In earlier studies effect of roughness on friction factor and velocity distribution was done by Nikuradse (18) for sand grain roughness. Later Cope (19) studied heat and momentum transfer for roughness elements followed by Nunner (20). Friction and heat transfer measurements for repeated rib roughnes in tube flow was done by Sams (21), Burnett(22), Koch(23). Webb, Eckert and Goldstein (24) conducted experiments using tube with internal fins and correlated their data in terms of roughness Reynolds number (Re^+) and roughness momentum transfer function $R(h^{+)}$.

Dipprey and Sabersky(25) analyzed their data in terms of roughness function for their experimental study. Sethumadhavan and Raja Rao(9) conducted experiments for heat and momentum transfer for the tubes with tightly fitted helical wire coils. Most of the works mentioned above utilized wall similarity concept and correlated their data in terms of roughness function and roughness Reynolds number by assuming two regions namely viscous region close to the wall of the tube and turbulent region which exists in the turbulent core away from the surface of tube. The same two-region flow assumed in the present study also. Due to the presence of disc across the flow in the tube generated nearly turbulent core flow at centre and viscous flow at the wall. Turbulence generated by the disc and swirl generated by twisted tape together presented intense turbulence that caused reduction of film thickness at the wall.

Correlation of data for the flow with insert promoters in terms of roughness function and roughnes Reynolds numbers is not found in literature. Despite this fact an attempt is made to correlate the data in terms of $R(h^+)$ and Re^+ using the following type of analysis. The basic similarity concept assumes the existence of two regions mentioned earlier ,namely, (i) the inner region near wall where the velocity distribution depends exclusively on the local conditions like y ,τ_w , μ, D_d, h and (ii) the outer region away from the wall where direct effect of viscosity on mean flow is negligible. For the inner region near the wall, the dimensionless velocity is given as

$u^+ = y^+$ ....(6)

where $u^+ = u/u^*$

$y^+ = y\ u^*/\nu$

For the outer wall region where the dependency of velocity distribution on molecular viscosity ceases to exist. The velocity distribution would follow the relationship

$u^+=1/k.\ln y^+ +c_1$ ....(7)

By the application of boundary conditions u=0,y=y0 where y0 is the thickness of laminar sub layer that depends on the turbulence generated , equation 7 reduces to

$u^+ =1/k \ln y/y_o$ ....(8)

The turbulence in the core and at the wall is significantly affected by the geometric parameters of the promoters employed in addition to the fluid velocity. In the present case, distance between tape and disc (h) and disc diameter (D_d) are the major characteristic geometric parameters as these are expected to be significantly effect the thickness of the laminar sub layer. In the present study the parameter h was chosen while computing u+

therefore, $y_o \propto h$ .... (9)

Equation 8 could be modified as

$(u_{max}-u)/u^* = 1/k \ln (y/h)$ ....(10)

Combination of equations 6 and 10 gives the velocity distribution equation for the turbulent dominated part of the wall region

$u^+=2.5\ln (y/h)+R(h^+)$ . ...(11)

The above equation presents modified velocity profile for the case of two regions in the presence of promoters. Assuming that equation 11 holds good for the entire cross section of the tube, the friction factor for the turbulent flow inside tubes with tape-disc assembly can be given the by integration of equation 11. The generated roughness function $R(h^+)$ is given by the following equation ,

$R(h^+)=2.5\ln[2h/d)]+\sqrt{(2/f)}+3.75$ ....(12)

Where $R(h^{+)}$ is roughness momentum transfer function The resulting format of equation for correlating the momentum transfer data with tape-disc assembly as promoter can now be written as

$R(h^+)=C_1(Re^+)^{b1}$ .... (13)

Here, C_1 is proportionality constant and b1 is an exponent Re^+ is roughness Reynolds number defined by the following equation

$Re^+ = (h/d).Re.\sqrt{(f/2)}$ ....(14)

where "h" is distance between tape and disc and 'd' is diameter of tube and f is friction factor. The data on momentum transfer was therefore similarly analyzed for tape - disc region. By using roughness function $R(h^+)$ in place of f and Roughness Reynolds number Re^+ in place of Re the following correlations was obtained by regression analysis. Correlation without incorporating dimensionless geometrical groups could be viewed hereunder,

Tape-disc region

$R(h^+)=0.601\ (Re^+)^{0.297}$ ....(15)

Average deviation =8.9, Standard deviation =11.0

The following correlation was obtained by incorporating dimensionless geometrical groups

Tape-disc region

$R(h^+)=3.573\ (Re^+)^{0.068}\ (\phi_1)^{0.050}\ (\phi_2)^{-0.033}\ (\phi_3)^{0.012}\ (\phi_4)^{0.228}\ (\phi_5)^{-0.037}$ ....(16)

Average deviation =8.1, Standard deviation =9.6

With the incorporation of geometric groups offered decrease in deviations.

Correlation plot for equation (16) was shown in figure 9.

CORRELATION OF MASS TRANSFER DATA

The effects of dynamic and geometric characteristics of the system on mass transfer could be correlated by relating Colburn JD factor and Reynolds number as suggested by earlier studies. A regression analysis of JD – Re data yielded the following equation for tape-disc region with large deviations indicating inaccurate fit.

$J_D = 1.8352Re^{-0.615}$ . ... (17)

Average deviation =41.52, Standard deviation = 64.90

By incorporating dimensionless geometrical groups, the following correlation was yielded

$J_D = 0.618Re^{-0.615} (\phi_1)^{0.885} (\phi_2)^{-0.041} (\phi_3)^{-0.133} (\phi_4)^{-0.003} (\phi_5)^{0.106}$ .... (18)

Average deviation = 35.66, Standard deviation = 53.18

In view of large deviation an alternative method of correlations had been considered by assuming similarity concepts. Following procedure outlined by Dipperey and Sabersky (25) for heat transfer, mass transfer similarity function could be derived. The derivation was similar to Sherwood(26) for smooth surfaces and the result was in the form of the following equation

$$(f/2.st)-1/\sqrt{f/2} = \int_0^{y_1^+} \left[\frac{1}{(1/Sc)+(\in_D/\nu)} - \frac{1}{1+(\in/\nu)} \right] dy^+ \quad .....(19)$$

The upper limit y_1^+ is an arbitrary value, large enough to be outside the region where molecular properties are important. The law of wall similarity shown by Nikuradse(18) to be valid for rough surface

$\in/\nu = f(y^+, Re^+)$ ....(20)

Similarly $\in_D/\nu = f(y^+, Re^+, Sc)$ ....(21)

By the incorporation of $\in/\nu$ and $\in_D/\nu$ in equation 19, the equation becomes,

$(\sqrt{f/2}st)-(1/\sqrt{f/2})+F == g^+(Re^+, Sc)$ ....(22)

For the constant F, the fully rough value of $R(h^+)$ has been added to both sides of the equation to make the g^+ function consistent with Dipprey's g(h) function for heat transfer. Therefore the above equation for mass transfer becomes

$g^+(Re^+,Sc) = [(\sqrt{f/2}st)-(1/\sqrt{f/2})]+R(h^+)$ ....(23)

Dawson and Trass (17) neglected $1/\sqrt{f/2}$ and $R(h^+)$ in mass transfer function while correlating their mass transfer data for high Schmidt numbers and arrived an expression for $g^+(Re^+,Sc)$ as

$g^+(Re^+,Sc)=St/St_o$ ....(24)

However in the present study due to wakes generated by disc and the swirl generated by the tape produced intense turbulence, which necessitated to retain $R(h^+)$ term in

$g^+(Re^+,Sc)$. Now the above equation will become,

$g^+(Re^+,Sc)= (St/St_o)+R(h^+)$ . ...(25)

The data on mass transfer was therefore analyzed by replacing J_D with $g(h^+)$ and Re with roughness Reynolds number (Re^+)

Where, $g(h^+) = (St/St_o)+R(h^+)$ ....(25)

$Re^+ = (h/d).Re.\sqrt{(f/2)}$ ....(14)

$R(h^+) = 2.5\ln(2h/d)+\sqrt{(2/f)}+3.75$ ....(12)

For tape – disc region , the following correlation was obtained by regression analysis.

Tape -disc region

$g(h^+) = 1.247\ (Re^+)^{0.243}$

Average deviation = 11.2, Standard deviation = 16.6

By incorporating dimensional groups and Schmidt number , the following correlation was obtained

Tape-disc region

$$g(h^+) = 343.95\ (Re^+)^{-0.176}\ (\phi_1)^{0.244}\ (\phi_2)^{-0.001}\ (\phi_3)^{-0.033}\ (\phi_4)^{0.402}\ (\phi_5)^{0.016}\ Sc^{-0.345} \quad ...(26)$$

Average deviation = 8.4, Standard deviation = 12.1

Correlation plot for equation (26) was shown in figure 10.

Dimensionless Groups:

$g(h^+)$ = Roughness mass transfer function = $(St/St_o)+R(h^+)$

J_D = Mass Transfer Factor $(k_L/V).Sc^{2/3}$

Re = Reynolds number = $dV\rho/\mu$

Re^+ = Roughness Reynolds number = $(h/d).Re.\sqrt{(f/2)}$

$R(h^+)$ = Roughness momentum transfer function = $2.5\ln(2h/d)+\sqrt{(2/f)}+3.75$

St = Stanton number = k_L/V

St_o = Stanton number for conduit without internals

Sc = Schmidt number $\mu/\rho D_L$

Sh = Sherwood number $k_L.d/D_L$

u^+ = dimensionless velocity, u/u^*

y^+ = dimensionless radial distance from the wall, $y\,u^*/\nu$

Nomenclature :

d = Diameter of test section, m

D_L = Diffusivity of reacting ion, m^2/sec

D_d = Disc diameter, m

E = Energy consumed using tape-disc assembly in the conduit ,N/m^2

E_o = Energy consumed for empty conduit,N/m^2

f = Friction factor ,$\Delta p\ d\ g_c/2LV^2\ \rho$

F = Faraday's constant = 96,500 coulombs/g-mol

g = Acceleration due to gravity, m/sec^2

g_c = Gravitational constant.

h = Distance between tape and disc, m

i_L = Limiting current, amp

k_L = Mass Transfer coefficient, m/s

k_o = Mass transfer coefficient of the empty conduit, m/s

L = Length of Test section, m

n = Number of electrons transferred

Q = Volumetric flow rate, m3/s

T_P = Pitch of tape, m/turn

T_W = Width of tape, m

T_L = Length of tape, m

u = Local velocity, m/s
u* = Friction velocity = $\sqrt{(\tau_w g_c / \rho)}$, m/s
V = Average velocity, m/s
y = Radial distance from the wall, m
$Y_1 = R(h^+)/ (\phi_1)^{0.050} (\phi_2)^{-0.033} (\phi_3)^{0.012} (\phi_4)^{0.228} (\phi_5)^{-0.037}$
$Y_4 = g(h^+)/ (\phi_1)^{0.244} (\phi_2)^{-0.001} (\phi_3)^{-0.033} (\phi_4)^{0.402} (\phi_5)^{0.016} Sc^{-0.345}$

Greek letters:
$\in$. = Eddy viscosity, m^2/s
$\in_D$ = Eddy diffusivity ,m^2/s
ϕ_1 = (T_L/d) = Dimensionless group
ϕ_2 = (T_W/d) = Dimensionless group
ϕ_3 = (T_P/d) = Dimensionless group
ϕ_4 = (h/d) = Dimensionless group
ϕ_5 = (D_d/d) = Dimensionless group
μ = Viscosity of fluid, Kg/m. sec
ν = Kinematic viscosity ,m^2/s
ρ_c = Density of manometer fluid, Kg/m^3
ρ . = Density of fluid, Kg/m^3
τ_w = Shear stress , N/m^2

References:

1. Colburn , A.P. and King , W.J.
Trans .A I C H E Journal 26, 196-206 (1931)
2. Turunen,I and Haario,H
Chemical Engg. Science, vol. 49,pp 5257-5269 (1994)
3. Prasad V.S.R.K.
"Studies on ionic mass transfer with impinging jets in closed cylindrical cells"
Ph.D. Thesis , Andhra University , Visakhapatnam ,India (1994)
4. Hwa Won Ryu, Young Soon Hyeon , Dong II Lee, Ho Nam Chang and O Ok Park,
Int. J. Heat and Mass Transfer, vol. 34,pp1909-1916 (1991)
5. Rong,X.Y., Kawaji,M. and Burgers,J.G.
A.I.CHE.Symp.Series, vol. 89, pp. 236-243 (1993)
6. Bergles,A.E.,
Progress in heat and mass transfer, vol. no. 1,p.331(1969)
7. Sitaraman, T.S
"Augmentation of mass transfer with coaxial string of spheres as internals
in tubes and fluidized beds"
Ph.D.Thesis , university of Madras ,Madras (1977)
8. Ventateswarulu, P and Jagannadha Raju, G.J.V.
4 'Th A.P.C. Ch.E, 87 Singapore, May ,13-15 (1987)
9. Sethumadhavan,R. and Raja Rao,M.
Int. J. Heat and Mass Transfer, vol. 26,pp1833-1845 (1983)
10. Sarveswara Rao, S.
Studies on Ionic Mass Transfer with co-axially placed cones on a rod in homogeneous

fluid and fluidized beds
Ph.D. Thesis , Andhra University, Visakhapatnam, (1983)
11. Sujatha, V.
" Studies on ionic mass transfer with coaxially placed helical tapes on a rod in homogeneous fluid and fluidized"
Ph.D. Thesis, Andhra University , Visakhapatnam, India. (1991)
12. Venkateswarlu,P.,
"Studies on ionic mass transfer with coaxially placed discs on a rod as turbulence promoter"
Ph.D., Thesis , Andhra university , Visakhapatnam ,India (1987)
13. Sujatha Kumari, B.
" Studies on mass transfer using coaxial – orifice turbulence promoters"
M.Tech. Thesis, Andhra University , Visakhapatnam, India. (2003)
14.Krishna,M.S., Jagannadha Raju, G.J.V. and Rao, C.V.
Ind. J. Technol. 4, 8 (1966)
15.Sudhakhara Rao, K., Rama Raju, C.V. and Jagannadha Raju, G.J.V.
Ind. J. Technol. 6, 129 (1966)
16. Bhaskar Sarma, C.
" Ionic mass transfer at circular and elliptical cylinders in cross flow of homogeneous fluid and in fluidized solids "
Ph.D. Thesis, Andhra University, Visakhapatnam (1978)
17. Dawson, David A. and Trass,Olev
Mass Transfer at rough surfaces
Int. J. of Heat and Mass Transfer, Vol.15,pp.1317-1336 (1971)
18 Nikuradse, J.
Laws of Flow in Rough Pipes. VDI FORSCH-Heft 361 (1933)
19. Cope, W. F.
The friction and heat-transmission coefficients of rough pipes,
Proc. Inst. Mech. Engrs., 145,99 (1941)
20. Nunner, W.
Heat transfer and pressure drop in rough tubes,
UKAEA Translation AERE Lib /Trans,786 (1956)
21. Sams, E.W.
Heat transfer and pressure drop characteristics of wire-coil type turbulence promoters,
USAEC Reactor Heat Transfer Conference (1956)
22. Burgoyne,I.B., Burnett,P. and Wilkie,D.
Forced convection heat transfer data from surfaces roughened by transverse ribs,
UKAEA,TRG Report 781 (W) (1964)
23. Koch,R.
VDI-Forscht, Series B,24,1 (1958)
24. Webb,R.L., Eckert,E.R.G. and Goldstein,R.J.
Int. J. Heat and Mass Transfer, vol. 14,pp601-617 (1971)
25. Dipprey,D.P. and Sabersky,R.H.
Int. J. Heat and Mass Transfer, vol. 6,pp329-353 (1963)
26. Sherwood,T.K.
Mass, heat and momentum transfer between phases,

Chem. Engg. Prog. Symp. Ser. No.25, 55,71-85 (1959)
27. Venkateswarlu,P.,
"Studies on ionic mass transfer with co axially placed discs on a rod as turbulence promoter"
Ph.D., Thesis , Andhra university , Waltair ,India (1987)
28. Changal Raju, D.
" Augmentation of mass transfer at the outer wall of concentric annuli in absence and presence of fluidizing solids -Effect of wires wound on central rod "
Ph.D. Thesis, Andhra University, Visakhapatnam, (1984)
29. Klaczac,A
Trans. A.S.M.E., J. Heat Transfer, 95, 557 (1973)

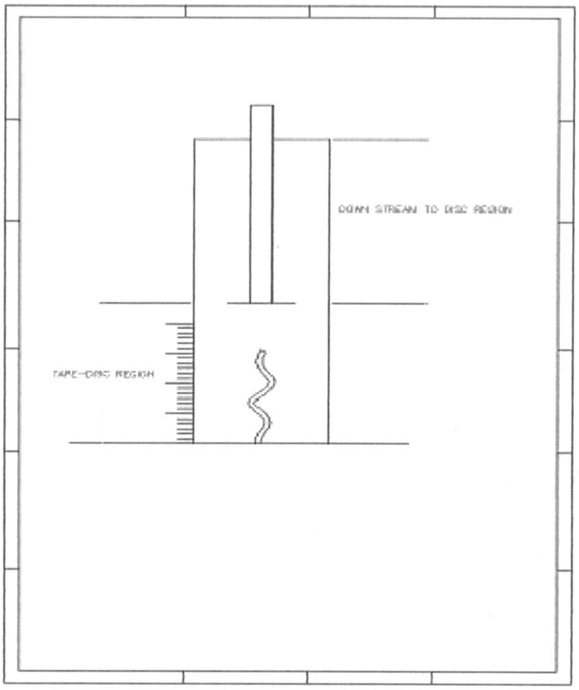

Figure 1

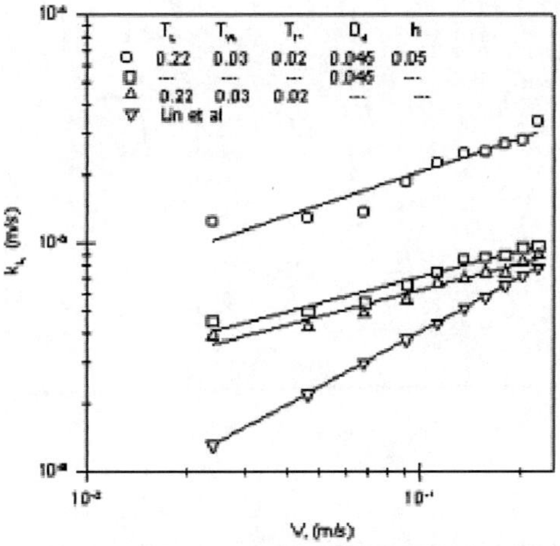

Variation of mass transfer coefficient with velocity
-for tape-disc assembly,single disc and single tape as turbulence promoters

Figure2

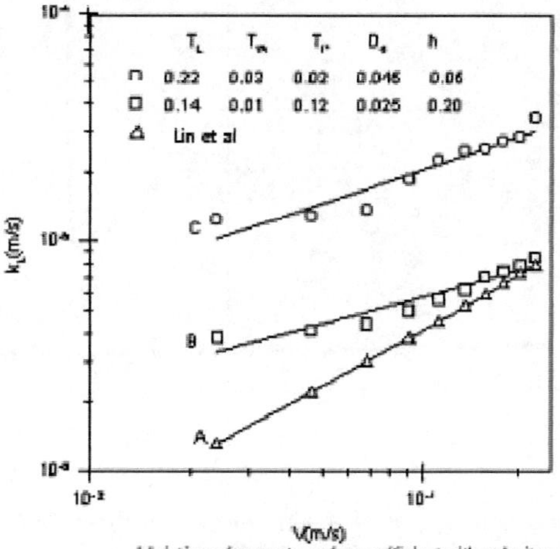

Variation of mass transfer coefficient with velocity
-for tape-disc region of tape-disc assembly as promoter

Figure3

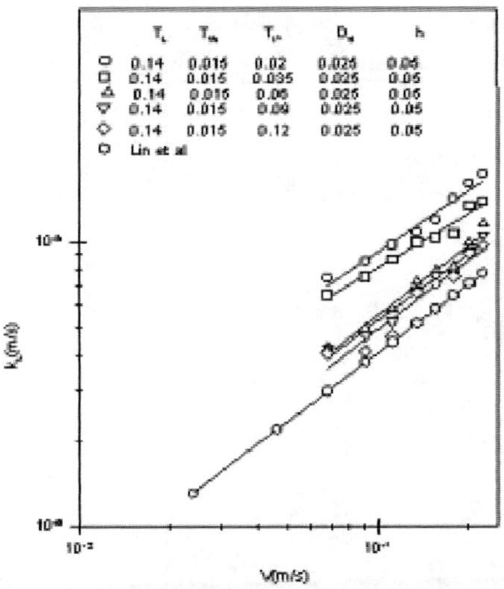

Vaiation of mass transfer coefficient with velocity
-Effect of pitch of tape in tape-disc assembly as turbulence promoter

Figure 4

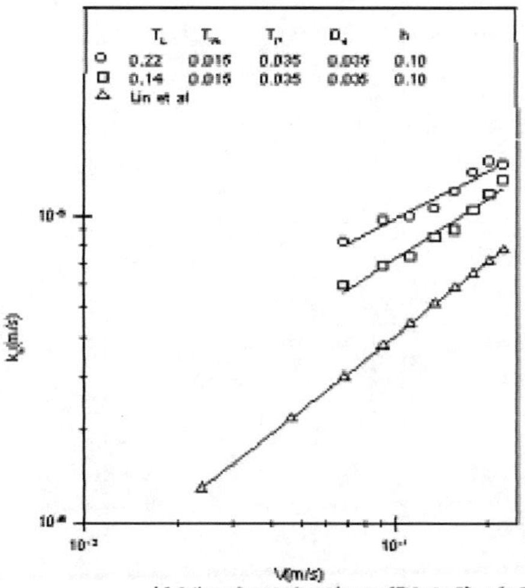

Variation of mass transfer coefficient with velocity
-Effect of length of tape for tape-disc assembly as turbulence promoter

Figure 5

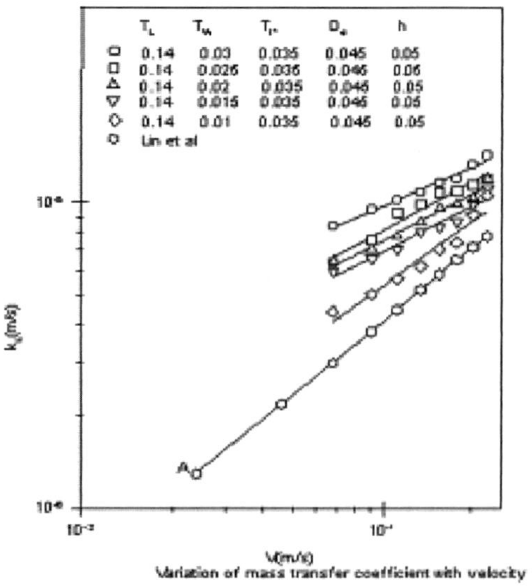

Variation of mass transfer coefficient with velocity
-Effect of width of tape for tape-disc assembly
as turbulence promoter

Figure 6

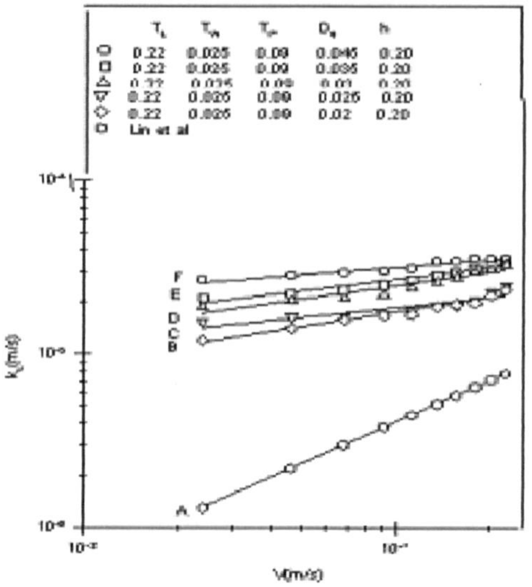

Variation of mass transfer coefficient with velocity
-Effect of disc diameter for tape-disc assembly
as turbulence promoter

Figure 7

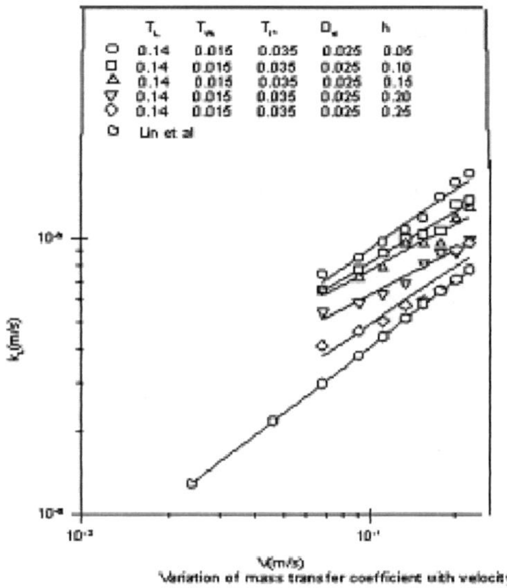

Variation of mass transfer coefficient with velocity
-Effect of distance between tape and disc
for tape-disc assembly as turbulence promoter

Figure 8

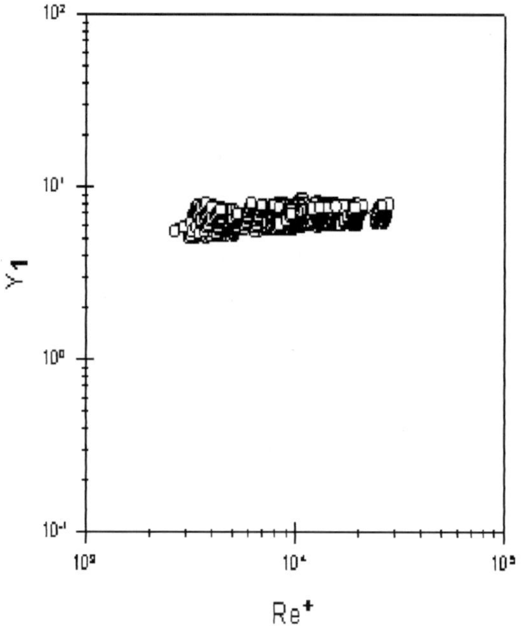

Figure 9

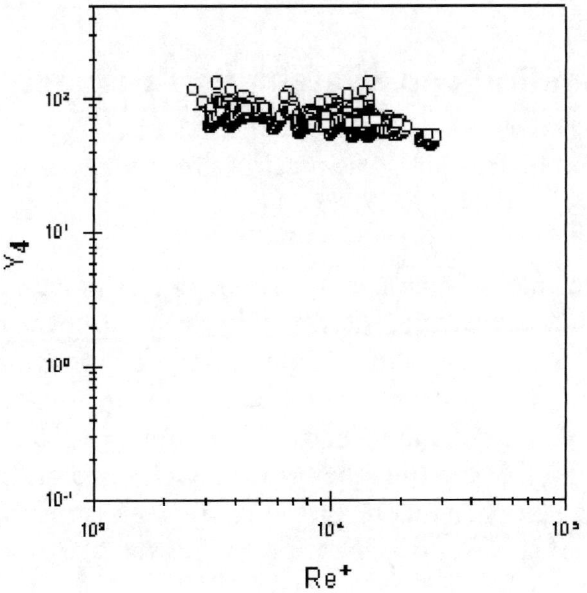

Figure 10

Process Intensification: New Understanding and Systematic Approach

Musaed Al-Thubaiti, Nasser Al-Azri, and Mahmoud El-Halwagi. Chemical Engineering, Texas A&M University, 942 WILLOW POND ST, COLLEGE STATION, TX 77845

This work introduces a new understanding of process intensification and its application through a systematic approach. First, we redefine the process intensification as any activity which is characterized by one or more of the following: 1) smaller equipment for a given throughput, 2) higher throughput for a given equipment size, 3) less inventory of certain material for the same production, 4) less usage of material utilities, energy utilities, or feed stock for a given throughput and given equipment size, 5) higher throughput through process for a given process size, and 6) higher process yield for a given process size. Then, the paper introduces a systematic approach that link local equipment intensification with other units in plant for optimal results. At the end, a case study is solved to illustrate the applicability of this approach and results are compared with literature

Experimental and Simulated Trajectories of Electrically Charged Drops in Non-Newtonian Liquid – Liquid Systems

Satya A. Gangu, Chemical and Petroleum Engineering, The University of Kansas, Learned Hall, 1530 W 15th Street, Lawrence, KS 66045, **Laurence R. Weatherley**, Department of Chemical and Petroluem Engineering, The University of Kansas, Learned Hall, 1530 W 15th Street, Lawrence, KS 66045, and Jerzy Petera, Process and Environmental Engineering, The Technical University of Lodz, ul Wolczanska - 175, 90-929 Lodz, Poland.

Process Intensification of solvent extraction processes for biological products, production of biodiesel from raw feedstocks, and of phase separations is an area for the potential small-scale applications of DC electric fields to liquid-liquid contacting. Simulated design of possible configurations of electrically enhanced reactors would be important in this context. The aim of this work was to validate a novel finite element algorithm for the prediction of three-dimensional trajectories for electrically charged drops. Earlier work had shown accurate prediction for newtonian liquid-liquid system and for a range of different contactor geometries. The current work sought to extend this to non-newtonian liquids. The results of trajectory studies of electrically charged non-Newtonian liquid drops dispersed into a non-Newtonian, immiscible continuous phase in an electrically enhanced rectangular glass column are presented here. Only the electrohydrodynamics of the droplets are studied and other modes of transport process are not considered. In this study, the continuous phase was selected on the basis of low electrical conductivity values ($< 10^{-10}$ S/m), a non-Newtonian rheology and optical transparency. The liquids used as the continuous phase were mixtures of polyisobutylene in mineral oil. Dispersed phases having a range of rheological properties were also studied. These included newtonian aqueous solutions, non-newtonian aqueous solutions of carboxymethyl cellulose (CMC) and aqueous Blanose. The motion of the droplets in three-dimensional space was videoed using a Sony digital camera. Image analysis software Image J was used to digitize the videos and to determine the time dependent xyz trajectory, the velocity data for the droplets and the corresponding drop sizes. The discrete drop and electrostatic dispersion regimes were studied. Physical parameters included, interfacial tension, viscosity, the density of both phases, and electrical conductivity of the dispersed phases. These were measured to physically explain the droplet behavior and also to model the system accurately. The effects of the applied D C electric field, the polarity of the electric field, the nozzle dimensions, the dispersed phase flow rate, the nozzle insertion depth, and inter electrode distance upon the trajectories will be presented. Some interesting phenomena associated with falling drops were observed. These included droplet breakup, coalescence, and oscillations occurring at higher DC electric fields. These effects were simulated with the finite element modeling. An initial good match between the experimental and simulated data was observed.

Keywords: electrically charged drops, D C electric field, non-newtonian, three dimensional trajectory, finite element analysis

The Use of a Controlled Cavitation Reactor for Process Intensification

Douglas G. Mancosky, Daniel A. Armstead, Bijan Kazem, Derek Parker, and Kelly Hudson. Hydro Dynamics, Inc., 8 Redmond Court, Rome, GA 30165

Hydro Dynamics, Inc. of Rome, GA utilizes an innovative next generation reactor technology based upon "controlled cavitation" called the ShockWave Power Reactor (SPR). The SPR works by taking a fluid, pure or impure, into the machine housing, where it is passed over the generator's spinning cylinder. The specific geometry of cavities in the cylinder and rotational speed create pressure differences within the liquid where tiny bubbles form and collapse. These collapsing bubbles generate shock waves that are used to heat and mix. These bubbles collapse only inside the cavities and away from the metal surfaces and are therefore not damaging to the equipment, the cavitation is controlled. For heating applications this means that a liquid can be heated with no heat transfer surfaces or temperature gradient, eliminating scaling and scorching. In the case of mixing, the shock waves increase the surface area of the compounds being mixed so that a higher mass transfer rate occurs. The technology can be used on liquid-liquid, liquid-solid and liquid-gas systems with gas to liquid ratios as high as five to one. Because of the process intensification, many traditional batch processes can be converted to continuous. Cavitation provides a unique mechanism of mixing not found in standard shear devices.

"Controlled cavitation" can be applied to any industry where increased mass transfer and process intensification are desired. The SPR can be utilized in the biodiesel transesterification reaction to provide high yields and quality in decreased reaction time. The SPR is also an excellent gas-liquid mixer and has been proven in laboratory and mill trials to greatly enhance oxidation. The technology also finds great utility in the manufacture of consumer products, as it easily mixes very thick and viscous fluids. Enhancements to the controlled cavitation technology also allow for the use of UV and electrochemistry with the process intensification.

Web Page: www.hydrodynamics.com

Microfiber Adsorbents for Improving Packed Bed Adsorbers

Alexios Harkiolakis, Joeri Denayer, and **Gino V. Baron**. Dept. Chemical Engineering, Vrije Universiteit Brussel, Pleinlaan 2, Brussel, B-1050, Belgium

Industrial gas or liquid phase adsorptive separations are usually performed as column processes, where the adsorbent takes the form of pellets packed in a column. To limit pressure drop, these particles have to be fairly large (1-5mm) and especially in the liquid phase, these then exhibit a strong mass transfer limitation by diffusion in the macropores or even large crystals in the particles. As a result, the breakthrough fronts are often very broad, even with a favourable equilibrium. The strong reduction of the particles to some 10 micron or the use at smaller scale operations of monolithic porous columns reduces the MTZ, but this can only be performed for short column heights like in high pressure chromatography in fine chemical or chiral separations. For large scale industrial processes (environment and petrochemicals) this is a much too expensive option. The only possibility is then to reduce flow rates and results in low volumetric productivities, large adsorbent inventory and column size. Especially when using a highly selective adsorbent to remove impurities from a large stream, this is economically not feasible. Usually, the adsorbent is only very partially loaded when breakthrough already occurs. The use of monoliths, supported small particles or thin layers on fibers with very open structures, where diffusion distances are very short, dramatically improve mass transfer and overall performance. On a volume basis, these materials however have a limited capacity and would require very large columns for bulk separations. Their cost is also prohibitive for large scale application. Recently, they have been proposed for use at the exit of a conventional packed bed that takes most of the load, but where a short bed of fibers or other finely structured material is used at the exit to catch the traces during early breakthrough. A limited capacity, but coupled to very fast mass transfer then allows to increase the breakthrough time by a factor 2 tot 3, and hence increase the bed loading by the same factor. These fibers or monolithic structures are ideal for the removal of traces of impurities from large streams, given their fast mass transfer and low pressure drops. In these applications, a large capacity is not needed and a thin layer on a carrier is sufficient(10-20% loading with 90% porosity materials). Such modified processes not only result in more compact equipment and lower investment, but also in reduced energy consumption in regeneration of the beds. Layers of adsorbents such as zeolites of 100nm can be produced on various supports showing virtually no mass transfer limitation. This paper presents a modeling study, identifying the range of parameters and applications where such a combined bed system might be of use and economically feasible. Preliminary experiments of adsorptive purification with metal fibers coated with zeolites will also be presented.

A Z STEP CASCADE PROCEDURE FOR EXTRACTION OF STYRENE FROM A MIXTURE OF STYRENE AND ETHYLBENZENE USING DIETHYLENE GLYCOL AS SOLVENT

Replace with logo

Dr. Kal Renganathan Sharma PE
Adjunct Professor, Department of Chemical Engineering (MC 110)
University of Illinois at Chicago, 202 Chemical Engineering Building
810 South Clinton Avenue, Chicago, IL 60607

Replace with logo

Abstract

Rather than the raffinate as feed to the subsequent stages, the extract is used as feed to the subsequent stage inorder to produce a product high in content of the extract. The stair-case combined graph procedure yields steps of Z for a multi-stage cross-current operation. This is different from the vertical rise steps in the design of the binary distillation column using the McCable Thiele method with constant overflow and underflow.

For example, a solution containing 50% ethylbenzene (A) and 50% styrene (C) need be separated at 25 0C at a rate of 1000 kg/h. The extract product composition at the end of the final stage is 92% styrene. Diethylene glycol is the solvent. It is added in a cross current fashion. The slope of –R/E at each stage was maintained a constant and equal to 1. The equilibrium data in solute free coordinates between the raffinate and extract phases are given in Figure 1.0. Rather than the composited extract the extract is removed at each stage and fed into the next stage. 10 ideal stages are required to achieve 92% styrene concentration in the final extract phase. Inorder to make a product rich in the solute the extract from each stage is given as a feed into the next stage.

Operating lines at each stage with a sloop of -1 are constructed starting with the feed composition of 0.5 at Y = X line. The extract from stage 1 is used as feed of stage 2 and so on and so forth. 10 ideal stages are required to achieve the target styrene composition. The composited raffinate composition can be calculated from the graph in Bancroft coordinates.Copy

Typical Extraction

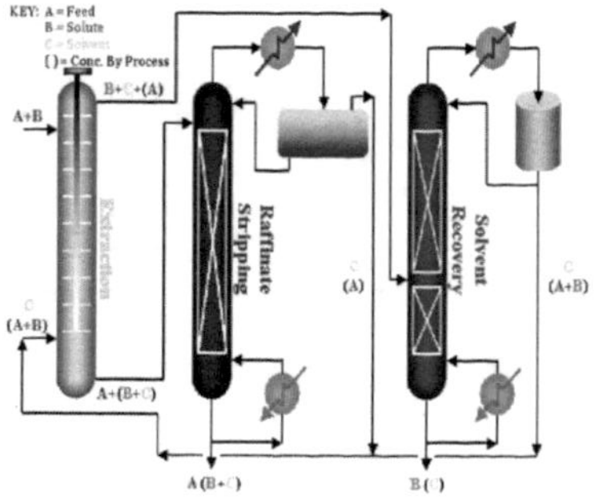

Triangular Phase Diagram

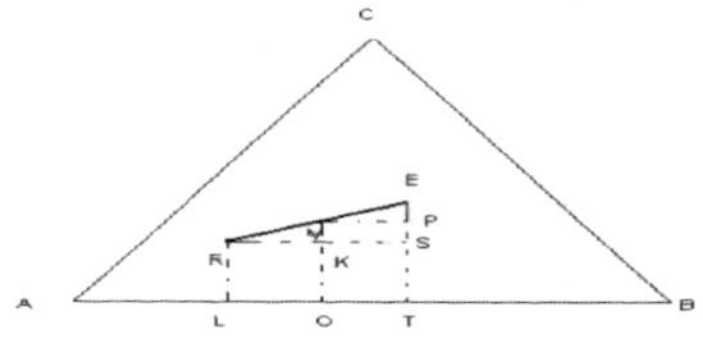

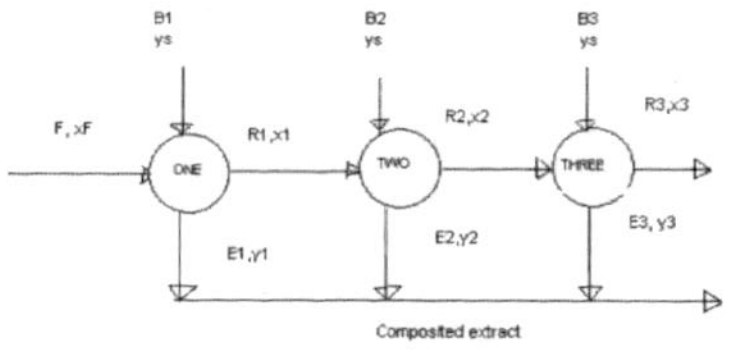

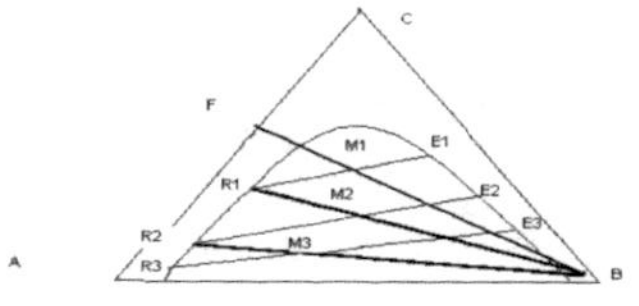

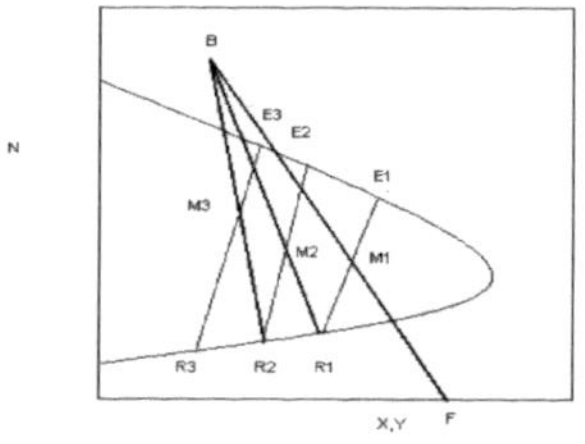

Combined Graph

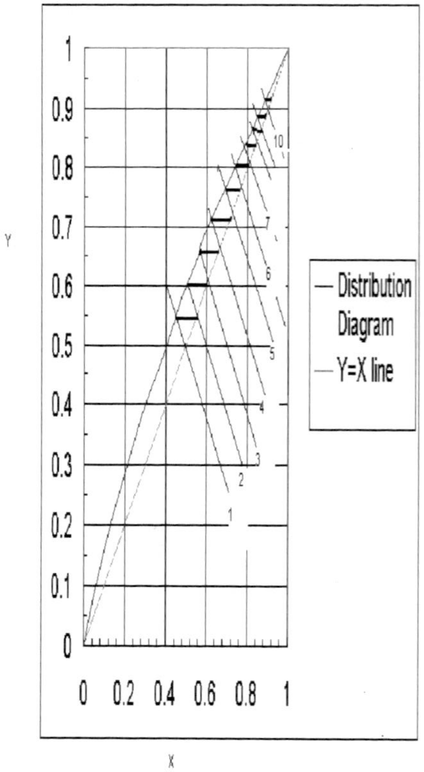

Etyhl Benzene/Styrene

A solution containing 50% ethylbenzene (A) and 50% styrene (C) need be separated at 25 0C at a rate of 1000 kg/h. The extract product composition at the end of the final stage is 92% styrene. Diethylene glycol is the solvent. It is added in a cross current fashion. The slope of –R/E at each stage was maintained a constant and equal to 1. The equilibrium data in solute free coordinates between the raffinate and extract phases are given in Figure 8.10. Rather than the composited extract as shown in Figure 8.8 the extract is removed at each stage and fed into the next stage. How many ideal stages are required to achieve 92% styrene concentration in the final extract phase. In order to make a product rich in the solute the extract from each stage is given as a feed into the next stage.

Discussion

Operating lines at each stage with a sloop of -1 are constructed starting with the feed composition of 0.5 at Y = X line. The extract from stage 1 is used as feed of stage 2 and so on and so forth. 10 ideal stages are required to achieve the target styrene composition. The composited raffinate composition can be calculated from the graph in Bancroft coordinates.

References

References

Belter, P. A., Cussler, E. L., and Hu, W.S., 1988, *Bioseparations*, Wiley, New York.

Catani, S. J., Vernon, N. M., Neiditch, D. S., Wiley, J. E. and Micinski, E. 2006, US Patent 7,049,435, *Extractive Methods for Purifying Sucralose*, Tate and Lyle, UK.

Cusler, E. L., 1997, *Diffusion : Mass Transfer in Fluid Systems*, Cambridge University Press, UK.

Inernet Website, Extraction Equipment, Koch Modular Process Systems, 2006, *Extraction Technology Group*, Paramus, NJ, USA

Kohr, W.J.,Shrader, V., and Johansson, C., 2005, US Patent 6802888, *High temperature heap bioleaching process*, Geobiotics, CO, USA.

Laddha, G. S., and Degaleesan, T. E., 1976, *Transport Phenomena in Liquid Extraction*, Tata-McGraw Hill, N.Delhi.

Lo, T. C., Baird, M. H. I., and Hanson, C., 1983, *Handbook of Solvent Extraction*, Wiley, New York.

Marsden, J.O., Brewer, R. E., Robertson, J. M., Hazen, W., Thompson, P. B., David, R., 2006, US Patent 7041152, *Method for processing elemental sulfur-bearing materials using high temperature pressure leaching*, Phelps Dodge Corp., AZ, USA

McCabe, W. L., Smith, J. C., and Harriott, P., 1993, Unit Operations of Chemical Engineering, Fifth Edition, McGraw Hill, New York.

Meinhold, J. F., Musto, J. A., Kramer, K. C. and Gottesman, M., (deceased), 1987, *Method for the decaffeination of roasted coffee extracts*, US patent 4,659,577, General Food Corp., White Plains, NY, USA.

Sharma, K. R., *Recycle of Uranium in Fission Products*, AIChE Spring National Meeting, New Orleans, LA, USA, March, 2002.

Sharma, K. R., 2005, *Damped Wave Transport and Relaxation*, Elsevier, Amsterdam.

Rao, G. and Sittig, M., 1997, *Dryden's Outlines of Chemical Technology – For the 21st Century*, East-West Press, N.Delhi.

Treybal, R. E., 1963, *Liquid Extraction*, McGraw Hill, New York.

Treybal, R. E., 1980, *Mass Transfer Operations*, Third Edition, McGraw Hill, New York.

Heat Exchanger Fouling Due to Catalysis

Hilario Garza, HODer, Shell Global Solutions, 3333 Highway 6 South, Houston, TX 77082 and Jonathan H. Worstell, HODER, Shell Global Solutions US, 3333 Highway 6 South, Room E-2061, Houston, TX 77082.

Catalysis, particularly polymerization catalysis, can foul heat transfer surfaces. Such fouling reduces the annual capacity of the production facility due to operating at reduced throughputs or due to downtime required for physically removing the fouling deposit. Such fouling is difficult to predict, as is the rate of such fouling. This paper describes a laboratory-scale performance test for determining the propensity of a catalyst to foul heat transfer surfaces and the methods used to quantify such fouling.

Tube Temperature Measurement by Vision

by Thomas Canty,P.E. and Keith DeMonstoy
JM CANTY INC
Buffalo, NY 14094

Abstract:

This presentation is intended to provide insight into non contact temperature analysis by vision and how it compares to other non contact technologies as well as traditional contact methods. Vision methods offer expanded instrument capacity as well as flexibility in monitoring the process due to the operator view of the process it provides and the many options available for data analysis.

Non contact means of temperature indication are primarily intended to provide relief from deterioration of traditional contact methods (i.e. thermocouple's etc…), costs of installation and maintenance. Optimum operation of Hydrogen furnaces, Ethylene furnaces and power boilers relies primarily on control of tube temperature to yield best energy efficiency and reliability of the process. It is also critical that hot spots be detected along the tube length in order to prevent an upset condition that will bring the boiler down for repair.

The basic theory is based on the radiative emissions of a body at temperature – as the body temperature increases, the radiation curve shifts towards the shorter wavelengths and that is why as a body heats up the eye cannot detect it until it reaches 1000F which is the approximate visible point where emitted radiation is dark red in color. As the temperature increases, the color graduates to bright red, then orange then toward white. A pyrometer functions by measuring a narrow band of wavelength intensity and comparing it to the intensity of a black body to determine the temperature. From the graph below it can be seen that if a single wavelength is monitored the intensity of that measurement changes as the body temperature changes (curve shifts) which then corresponds to a different temperature on the calibration scale.

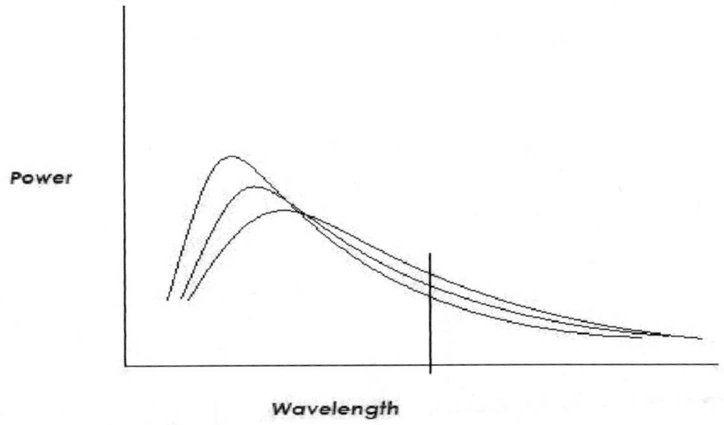

The same is true of a camera system measuring the increasing intensity of an object as it radiates from approx 1000 F (dim, dark red color) on up the temperature scale to 3000+ F for instance which would be a bright white. From the graph above one can see that as color goes from red toward white the peak intensity is always increasing and it is the measure of this intensity which indicates the object temperature. Using an RGB camera the measurement can also be made by monitoring the individual color components of the signal impinging on the chip. For industrial uses, the measurement of intensity usually suffices.

Traditional Technology

Standard instrumentation used in industry include thermocouple type devices (contact) and thermal imaging and pyrometer systems (non contact). In severe temperatures and atmospheres, the contact devices deteriorate quickly and provide gradually less reliable information leaving the operator running blind. They also create a discontinuity on the surface they are welded to which, in the case of a furnace tube, can lead to premature tube failure.

Non contact systems have difficulties with calibration drift and emissivity adjustments to account for the fact that many items being measured are not black body emitters. In addition, these technologies are single point measurement systems and are best used as hand held devices which can be difficult to use and provide subjective measurement results based on operator preferences for targeting.

Vision Technology

Vision technology offers several benefits these other technologies do not. The visual aspect of the system allows the operator to view the process and confirm the readings received. It also allows flexibility to set up measurement zones on screen at any time to assess any or all areas of the process in view should a change be required. Since vision systems are generally fixed installations, an added benefit is that the information is taken from the same process points and features within a given measurement zone at all times which means there is no error in sighting of the instrument.

The vision system is also unique in that it does not suffer calibration drift as almost all instruments do. Once the calibration is set the readings repeat themselves within a couple degrees F. The calibration is based on thermocouple or pyrometer readings initially, however after that calibration is finished it does not have to be redone in the future as long as the system remains in it's location.

Process history and performance can also be best documented by a vision system. First off, the camera can analyze any area of process that is within the systems view. This affords the operator the capability of viewing many points of interest instead of just single point temperature readings. As an example, if the system is monitoring a tube bank it can assess the temperature of the tubes for the full length of view. If there is a local flow problem in a tube it will become overheated in one section. The camera can locate that section and indicate it's temperature rise over the rest of the tube or adjacent tubes. This information allows the tube location to be identified as well as the position in the tube that has encountered the upset condition. Over time this historical data, recorded in video or still image format by the control software, may be significant in improving the process. In addition to the ability to monitor the full area of the view, the full area of view can also be segmented into multiple zones of measurement, i.e. each tube having it's own measurement zone for automated process control capabilities. This capability allows the tube to be operated at optimum efficiency since the addition of excess heat can be avoided thereby reducing energy draw to operate the furnace.

System Demonstration
Run Video Here - http://www.jmcanty.com/video.html

Following are typical images of a metal rod heated in the near IR range and into the visible range. The first image shows the rod at 850 F which is imaged by the IR capabilities of the system. Also shown on this image are typical measurement zones. As can be seen, these can have any size or direction at the will of the operator and allows measurement of any area in the field of view. The second image shows the rod as it reaches 1000 F. Obviously the higher intensity radiation in the visual range creates an appropriate response by the CCD chip.

Image of element at 850 F

Image of Element at 1000 F

Performance on Tube Array

The following shows an image of a tube array with four measurement zones configured on 4 of the tubes. The measurement readings are indicated in the table at the bottom of the image and are listed in gray scale values. All measurements are within two percent of each other.

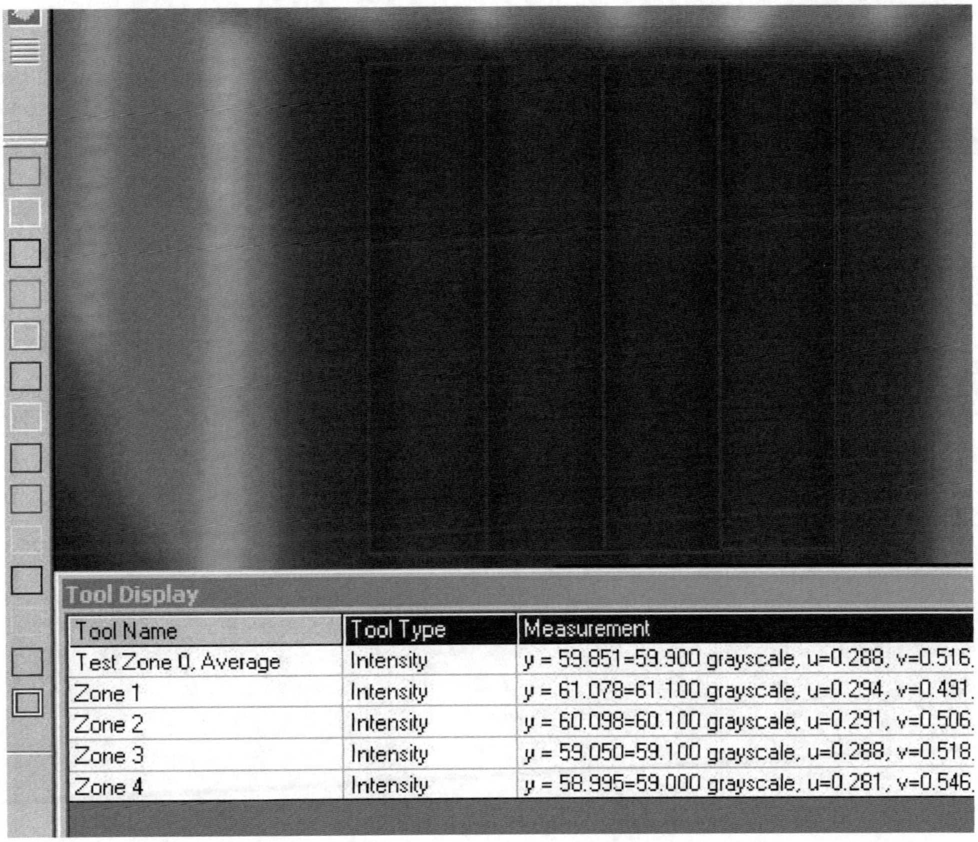

As thermal loads are fluctuated up and down the following graphs track those changes. The cooling cycles are left to natural convection and thus are not linear in nature. For comparison, a second heating cycle is included to demonstrate the repeatability of the vision system which meets the testing requirements of ASTM E29 for calibration and accuracy of temperature measurement devices.

For the following curves, the vertical scale is gray scale intensity which is measured 0 – 255 (pitch black to bright white). The hotter the tube, the higher, or more intense, the gray scale reading. This value in process would be calibrated to the temperature. The horizontal scale is the image number, or time.

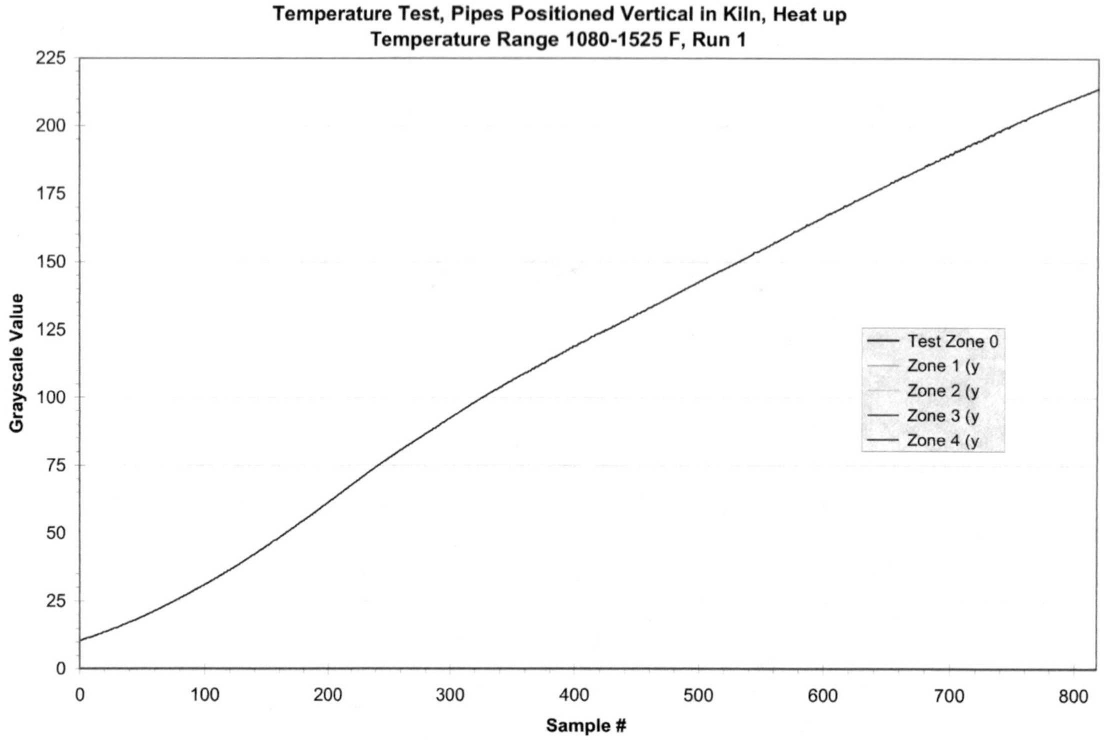

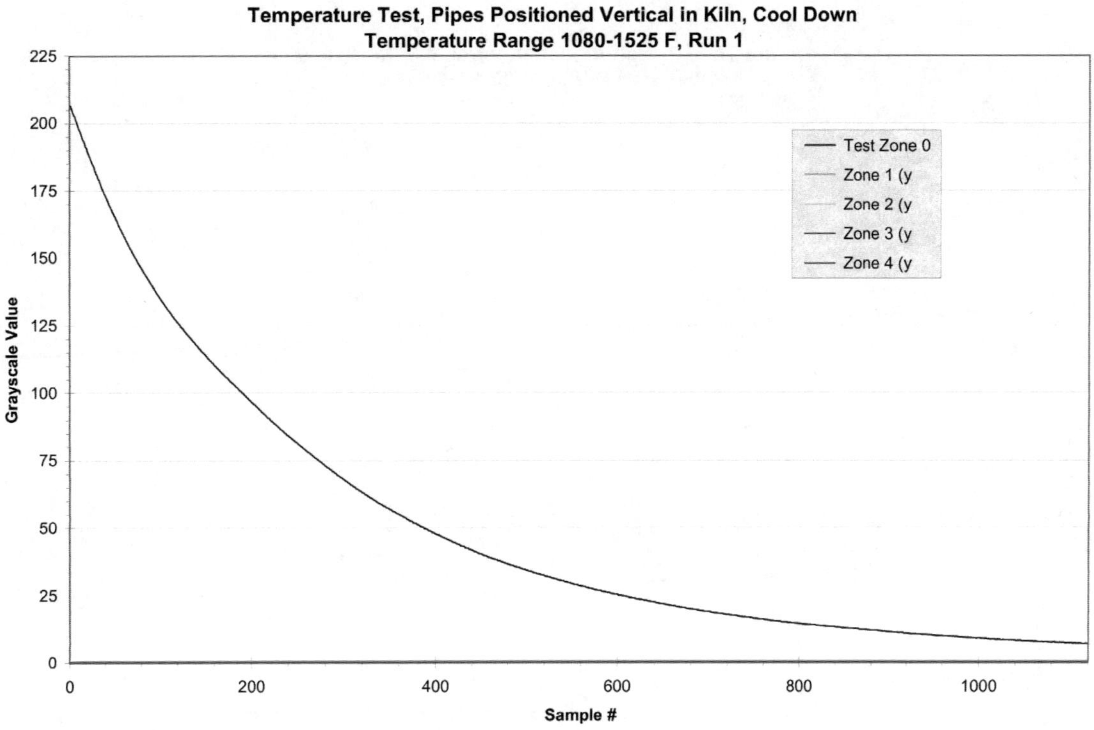
Temperature Test, Pipes Positioned Vertical in Kiln, Cool Down
Temperature Range 1080-1525 F, Run 1
Grayscale Value
225
200
175
150
125
100
75
50
25
0
0
200
400
600
800
1000
Sample #
Test Zone 0
Zone 1 (y
Zone 2 (y
Zone 3 (y
Zone 4 (y

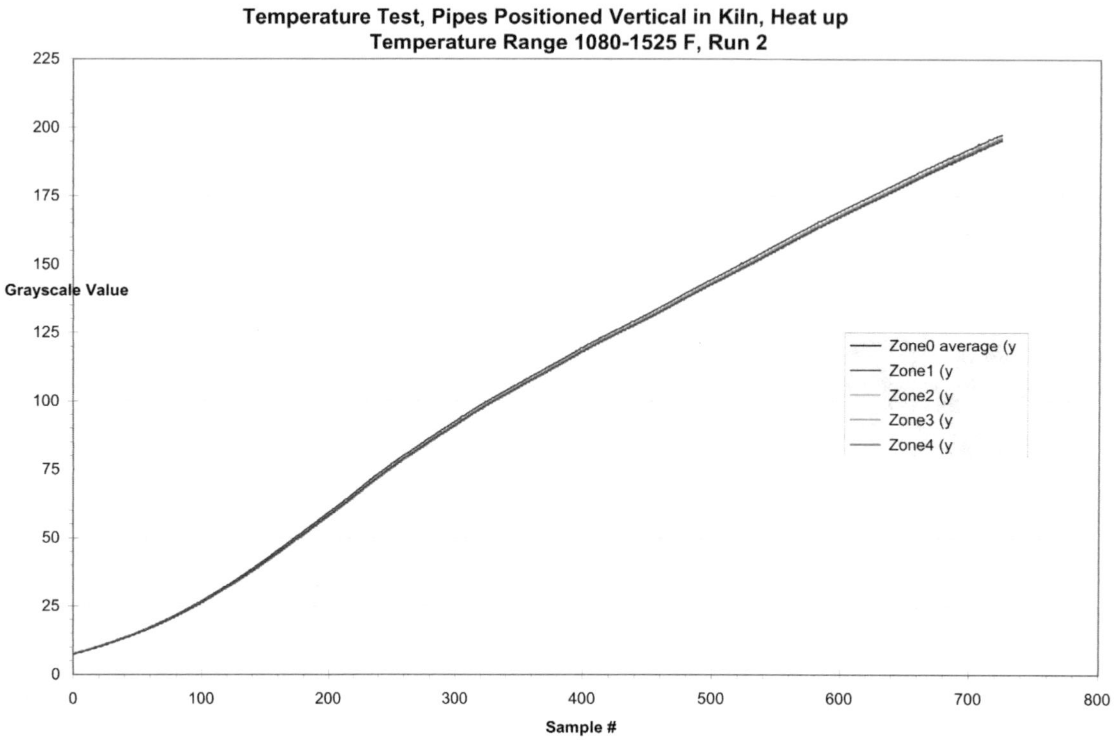

System Components:

The equipment that comprises the vision system required includes a High Temperature Camera and a computer/software package designed to analyze temperature. The camera must be able to survive high temperature environments and maintain a clean lens system. Depending on fuel source and purity of the furnace atmosphere, air for cleaning may or may not be required. Canty's design places a protective quartz shield in front of the lens system and conducts air for cleaning past that lens which prevents foreign particulate from contacting and sticking to the quartz shield. Should air be lost, the shield may coat over, however that can be removed and replaced with another – the lens components are unharmed.

Typical Camera - http://www.jmcanty.com/overview/P.Process%20Vision/A7355.PDF

The power of the software is critical in performance of the system. Having the analysis capability alone may not be enough as typical high temperature processes often exhibit viewing difficulties that can cause for false readings. The ability of the software to "intelligently" analyze the images is required to get good results. For instance, if there is smoke in a process view occasionally it can falsely lower the temperature readings. The software must be able to recognize this and discount those readings.

Challenging the Way Design Margins Are Added to Heat Exchangers

David King, Operations Efficiency, BP, Chertsey Road,, Sunbury-on-Thames,, Middlesex, United Kingdom

The traditional method of sizing heat exchangers often leads to unnecessarily large design margins that increase capital cost, footprint & weight and lead to poor operating efficiency. An alternative approach is proposed that is based on tackling the root causes of loss of performance and uses a revised design methodology to mitigate against fouling.

Fouling Factors: What Are They --- How to Cope with Them

Shawna S. Berg[1], Jonathan H. Worstell[2], and Richard E. Robertson[1]. (1) Geismar Manufacturing Facility, Shell Chemical Company, River Road, Geismar, LA 70734, (2) HODER, Shell Global Solutions US, 3333 Highway 6 South, Room E-2061, Houston, TX 77082

Many chemical manufacturing facilities and most chemical reactors are heat transfer rate limited. In other words, the annual capacity of the production facility is limited by the rate at which heat is added or removed from the chemical process. The addition or removal of heat is done via process heat exchangers. Unfortunately, heat exchanger performance deteriorates with time. Deposits on the walls of heat exchangers cause this performance deterioration. The resistance to heat transfer due to fouled heat exchanger surfaces can be monitored and remedial action taken to restore heat exchanger performance if the proper calculations are made. This paper outlines how to monitor heat exchanger performance and what can be done to restore heat exchanger efficiency, particularly when fouled on the process-side of the exchanger.

Polymer Fluid Properties and Heat Exchanger Performance

Harvey E. Atwood, RD&T, Kraton Polymers LLC, 3333 Highway 6 South, Westhollow Technology Center, Houston, TX 77082

This paper discusses heat exchanger performance as a function of polymer fluid properties. In such cases, heat exchanger fouling acquires a molecular aspect. An example of such fouling will be presented during this talk.

An Integral Technique of Monitoring Localized Coking in Refinery Fired Heaters

C.B. Panchal[1], Steven A. Lottes[2], and Michael Petrick[2]. (1) Argonne National Laboratory, 9700 S. Cass Avenue, Argonne, IL 60439, (2) Energy Systems Division, Argonne National Laboratory, 9700 S Cass Ave, Argonne, IL 60439

A typical refinery consumes about 445,000 Btu/barrel of crude processed, most of which is provided by process heaters. The recent trend of high utilization in the refining industry enforces operating of process heaters close to, or beyond, design capacities. Argonne is pursuing a long-term project for developing a pipe-flow two-phase flow process model, with the objective of predicting localized coking in straight and in U-bend regions. The threshold fouling developed for crude oil is incorporated into the process model to determine localized coking in crude and resid refinery heaters. The prediction results show localized variations of coking in straight and U-bend regions. An integral technique consists of this fired heater model and high temperature fouling/coking unit to monitor change in coking trend of the feedstock. The coking data from the fouling/coking unit are feedback into the prediction model. This integral technique combined with discrete as well as infra-red measurements of the tube metal temperatures will provide a real-time monitoring of the fired heater.

To Toll or Not to Toll: That Is the Question

Jonathan Worstell, HODER, Shell Global Solutions US, 3333 Highway 6 South, Room E-2061, Houston, TX 77082

Globalization requires all chemical manufacturers to reduce their product development cycle time. One method for reducing product development cycle time is to use contract manufacturers; ie, tollers, for market development and supply. This paper discusses how to decide whether to use a toller for product development and when to start building and operating an in-house manufacturing facility.

Waste Management: Ignorance Is Not a Winning Defense

Marty J. Pierce, Product Safety and Regulatory, Shell Global Solutions, One Shell Plaza, Houston, TX 77082 and Jonathan H. Worstell, HODER, Shell Global Solutions US, 3333 Highway 6 South, Room E-2061, Houston, TX 77082.

Many chemical manufacturers and chemical marketers are using contract manufacturers; ie, tollers, to reduce product development cycle time. Such companies expend considerable effort to ensure safe operation of the process at a toll facility and to ensure production of quality product by the toll facility. Waste management and waste disposal are sometimes lost in the rush to produce a new product. However, waste disposal should be the second highest item on the "to do" list for a potential tolling project. Why? Because people go to jail for mishandling waste. And, ignorance of waste disposal law is not a viable legal defense. This paper outlines the issues involved with managing waste generated by a new chemical product at a toll facility.

Using Tollers for Chemical Product Manufacture: the Details of Success

Mike J. Doll, HODer, Shell Global Solutions, 3333 Highway 6 South, Houston, TX 77082 and Jonathan H. Worstell, HODER, Shell Global Solutions US, 3333 Highway 6 South, Room E-2061, Houston, TX 77082.

Globalization requires chemical manufacturers to reduce their product development cycle time. One method for reducing product development cycle time is to employ contract manufacturers; ie, tollers. However, producing product at a toll facility is fraught with potential --- for failure. It requires attention to detail and above average communication skills to produce quality product at a toll facility. This paper outlines those areas requiring attention to detail and it emphasizes the need for thorough, clear, concise communication during a tolling project.

Essentials for Success with Contract and Toll Manufacturing

John M. Gohndrone, Dow Corning Corporation, 2200 W. Salzburg Road, PO Box 994, Midland, MI 48686-0994 and Rick J. Surma, Procurement, Dow Corning Corporation, 2200 W. Salzburg Road, PO Box 994, Midland, MI 48686-0994.

This paper will discuss the process that Dow Corning uses in selecting and working with contract manufacturers. Specific topics to be covered include when and why to consider contract manufacturing, how to select a contract manufacturer, the risks involved, and the typical workflow that is used at Dow Corning. The presentation will include examples to highlight the advantages and disadvantages of working with contract manufacturers.

Pilot Plant Design and Construction: Benefits of Outsourcing to a Pilot Plant Specialist

David Edwards, Sales and Marketing Department, Zeton Inc., 740 Oval Court, Burlington, ON L7L 6A9, Canada

Once the decision to pilot has been made, the search begins for existing facilities, both within the company and possibly externally, suitable for completing this work. Should a suitable fit not be identified, a new pilot plant will need to be designed and built.

At this stage, the decision then becomes whether to design and build the pilot plant in-house, or to outsource. This presentation will explore the benefits of outsourcing pilot plant design and construction to a pilot plant specialist.

Web Page: www.zeton.com

Session 12D05

METHODOLOGY FOR EVALUATING TOLL MANUFACTURERS

BY

Osman Aboul-Nasr
Chemical Technology Center
Fluor
Greenville, SC 29607-2762

Prepared for Presentation at AIChE Annual Spring Meeting
Houston, TX
April, 2007

Session 12D05
To Toll or Not to Toll: That Is the Question

"Unpublished"

METHODOLOGY FOR EVALUATING TOLL MANUFACTURERS

ABSTRACT

Toll manufacturing can provide significant time and capital cost savings, if a qualified toll manufacturer is chosen. The proper methodology is essential to evaluate if they can perform effectively and reliably, based on quantifiable measures, past performance and present capabilities. The issues and concepts addressed in the evaluation of tolling facilities are addressed here. Using a weighting methodology, the compliance of all manufacturers is then evaluated and compared.

This proprietary approach can help the owner in evaluating in-house production versus tolling. It can provide guidelines in developing a complete evaluation procedure for a specific purpose, or for general application. This methodology can help the toller in developing the documentation and procedures necessary to comply with the owner's requirements.

Fluor can help align the parties together and facilitate the implementation of a successful toll manufacturing process.

The following is an outline of the presentation, and highlights the approach. Many details were omitted from this manuscript. ***Not all the information is provided in this ABSTRACT. Please request a copy of the complete presentation from the author.*** Fluor will help the owner develop a complete evaluation procedure for a specific purpose, or for general application. Fluor can also help the Toll manufacturing in developing the documentation necessary to comply with the owner's requests. Fluor can help bring the parties together and facilitate the implementation of a successful Toll manufacturing enterprise. Fluor does not assume any responsibility for the implementation of the procedures listed below, unless approved by Fluor.

Tolling Decision

Toll manufacturing is a means to expedite marketing and manufacture. Toll manufacturing has traditionally been used in marketing studies and small-scale specialty products. Fast market entry is an important for commercial success, giving an edge to toll manufacturing. It also reduces capital expenditure, operator training, etc. On the other, technology owners may stand to lose confidentiality, quick response...

Product cost can be assessed easily by competitive bidding. However, the cost of manufacturing the product, from production to delivery and end use, is influenced by many factors and is not assessed easily. An economic evaluation should play a significant role in the final selection. The choice of the proper Toll manufacturing could mean the difference between success and failure for the venture.

In this paper, we will assume that the decision to Toll manufacturing has been made, and will present an approach to selecting the optimum Toll manufacturing for the purpose.

How to Choose

As in any comparative study, the owner needs written procedures that provide a systematic means of verifying the performance of the Toll manufacturing. Evaluation should address the performance of the process system, the operator training, and the safety procedures. The complexity of the evaluation should be proportional to the overall value of the project/product. The qualification of each system should follow a standard protocol. It will include the objectives and acceptance criteria, developed independent of the opportunity at hand.

Toll manufacturing evaluation provides a valuable protection against costly capital mistakes. A professionally executed evaluation program is essential for a successful project that provides the client with a cost-effective Toll manufacturing that meets internal in-house standards and requirements. Fluor can help the client to develop an effective approach, tailor it to fit the owner's needs, as well as providing the means necessary to implement it.

EVALUATION METHODOLOGY:

The owner should clearly determine their needs. The following punch list should serve as a guide in choosing the optimum Toll manufacturing. Each section should be given a certain weight, and a minimum threshold of acceptance. The weight given to each section or item is at the discretion of the owner and can change with product.

Fluor has developed proprietary software to help the owner prepare for the selection process. It will also help the Toll manufacturing get ready to satisfy the stringent requirements of the most discerning technology owner. The owner should view them as a clear road map for a successful venture. The Toll manufacturer should treat them as the first in many successful cooperative endeavours.

A. Formal Documents.

Setting up the rules for confidentiality, timing, business relationships, etc., reduces the potential for delays and misunderstandings later. Toll manufacturing agreements are more than pricing agreements; they should be the basis for a win/win partnership.

Tolling in foreign countries may present special considerations, and the laws of both the USA and the other country must be followed.

B. Process Definition

This owner must balance the confidentiality of the information with the clarity of the goal. A partial list may of the information.

- Process description, defining the chemistry and operating conditions
- Raw materials, complete with MSDS and special requirements.
- PFD's, mass and energy balances
- The type of unit operation, and the preferred and unacceptable types of equipment and materials of construction.
- Potential and possible releases, pollutants, toxic and hazardous wastes.
- Potential situations for producing uncontrolled kinetic releases, runaway reactions, unacceptable by-products, and how to avoid them.
- The consequences of unusual operations.

- The extreme temperature and pressure from the worst scenario.
- Develop written procedures for operating under these conditions.
- The required utilities, their level and flow.

C. Defining the Equipment.

The following information should be available to the owner, to ensure that the toller has the capabilities to perform the process successfully.

- Pressure vessel certificates and documentation.
- Basic layout drawings and safety features
- What are the installed relief valves/rupture discs? Additional details will be required.
- Heating/cooling capabilities, media and proper relief devices
- Controls and instrumentation critical to the safety of the process, duplicate/redundant systems.
- Applicable safety/industry/government/insurance standards
- Explosion-proof rating of the location
- Two accessible emergency exits

D. Maintenance Practices.

Checking the maintenance practices of the toller will help to differentiate between good and average manufacturers. Safety is paramount. The toller should provide the owner with the following typical information. More information is available form the author.

- Relief valve testing and inspection schedule
- Procedure to control "hot work" operations
- Systems control vessel entries
- Vessel inspection program
- Procedure to de-pressurize pressurized lines
- Which critical valves are identified and checked on a regular basis?
- Type of emergency repair equipment
- Certificates for fired and unfired pressure vessels
- Is there a program to calibrate process control instruments?
- Is the emergency alarm systems adequately maintained and tested?
- Are there steps taken to assure that auxiliary emergency equipment are adequately maintained and working properly?

E. Identify the Utilities.

The toller should provide the following typical requirements satisfactorily. More requirements are shown in the presentation and available form the author

- List of utilities available at facility and their source: local or purchased.
- List of utilities critical to safe operations
- The locations of emergency shut-offs for natural gas lines in the event of a gas main breakage?

F. Safety and Industrial Hygiene Practices

The owner should clarify the ownership issues of the waste materials as dictated by law. The toller should be able to address the following typical points. A more comprehensive list is available in the presentation material and from the author

- The qualification of the person responsible for the safety and industrial hygiene
- The process of reviewing all operating process procedures from a safety, industrial hygiene and environmental impact perspective before modifying procedures.
- What programs are used to control and check the safety and health programs of outside contractors?

G. Quality Control Procedures and Practices

The client should determine the capabilities of the toller to carry out standard testing and quality control procedures.

- How are raw materials and products identified prior to transferring to bulk tanks?
- What system is used to assure that all products have been properly checked, packaged and marked prior to shipment to the customer?
- Will there be an on-site QC representative of the client?
- What is the level of ISO certification? Is it commensurate with the product requirements?
- Are there any outside labs nearby that are required to do additional work: certification, research, university or other institutions?

H. Emergency Procedures.

Emergency procedures may have to be modified to accommodate the client's special requirements. This is an important aspect of any manufacturing process.

- Is there a written emergency procedure for the process and the entire facility
- Is it clear in the emergency plan who is in charge, and his individual roles
- Are all employees trained in emergency procedures?
- What are the possible emergencies that have been considered?
- What outside services have been incorporated into the emergency plan?
- What pre-arrangements have been made with these outside services?
- What outside service people have visited the plant?
- Are duties clearly defined and provisions made for: fire fighting, terrorist activities, security safety, medical utilities, outside assistance, media contact, evacuation of facility and of surrounding area), civil disobedience ...etc
- What system is in place to update emergency procedures? (There are a few questions to be asked about updating practices)
- The frequency and effectiveness of emergency drills
- Lists of what is considered to be critical documents such as plot plans, chemical inventories, etc.

I. Storage Facilities

The owner should identify the capabilities of the toller as well as potential incompatibility in case of spill. Typical requirements are listed below. A more exhaustive list is available from the author.

- Is all bulk chemical storage inventoried?
- A list of hazardous materials and their inventories. Could it be reduced?
- A list of storage tanks, their capacity and contents and location on a plot plan.
- Are all tanks provided with emergency vents?
- What types of vapor collection systems are provided for tanks of volatile, toxic, or odoriferous, etc. materials?
- What is the worst tank failure that could be experienced at your facility?
- What procedures are used to assure that incompatible materials are stored separately (acids, flammables, oxidizers, corrosives, etc.)?
- What inventoried materials in drums or tanks are capable of auto-polymerization?

J. Distribution and Transportation

The owner has to decide the role of the toller in distribution. Close attention has to be paid to federal, state and local laws and regulations governing the ownership of the product and hazardous materials. Typical questions to be investigated are, but not limited to, the following

- Are tank cars and trucks the appropriate type prior to loading?
- Is there a procedure for inspecting tank cars/trucks inspected prior to loading?
- Are "outside" truckers (not employees) permitted to unload material into tanks without supervision?
- What procedures assure error-free unloading and loading at your facilities as well as at customer's facilities when delivered in bulk?
- What procedures are in place to respond to transportation accidents that occur for incoming raw materials and your shipped products?

K. Training Procedures

The attention of the toller to training is a sign of their dedication to quality and safety. This section should be considered in addition to the specific training required for operating and maintaining the facilities.

- What techniques are used to determine the information necessary for employees to do their job, provide this information and evaluate employee's performance?
- What are the procedures taken to inform employees of the hazardous properties of the material?
- What limitations are placed upon the activity of employees that have not completed all the training programs necessary for the safe performance of the job?
- Are there any special additional training requirements to handle the materials and equipment needed for the new product?

L. Security

Several factors are needed to prevent industrial espionage and maintain confidentiality on the operators' level. The importance of this aspect is often underestimated. Security is also important, especially at high terrorism alert levels. Fluor can develop a detailed security analysis, based on our diversified national and international experience.

Special attention is necessary to keep up with the regulation governing the source and destination of material used in the process.

M. Offshore Toll manufacturing

Although toll manufacturing is generally performed within the US, the same principles apply to offshore facilities, with the additional consideration of the costs and security aspects. An economic assessment should be performed to substantiate the viability of offshore tolling. Security and confidentiality will deserve additional stress, and Fluor, through our international expertise and experience, can provide valuable assistance in this area. Fluor would consider providing a detailed security analysis for most of the countries where Toll manufacturing is a potential alternative.

Software

Fluor has developed proprietary software that can help both the owner evaluate the and the Toll manufacturing get ready for a cooperative venture. By facilitating the process, Fluor can bring assist the owner, the Toll manufacturing or both in taking the proper steps towards ensuring the success of the venture. The software starts with the above principles, extends them to include additional criteria, and develops a matrix that can compare Toll manufacturing facilities and give the proper weights assigned in partnership with the owners.

Conclusions

By developing a Toll manufacturing selection process, the technology owner will ready to launch a tolling venture that eliminates most of the unforeseen pitfalls. The choice of the proper Toll manufacturing can mean the difference between success and failure. For the Toll manufacturing, getting ready for a rigorous analysis can mean more than just the job at hand, but enhanced reputation, satisfied clients and higher potential for future business.

The above is an extended abstract of the methodology, please contact the author for further information/

Dr. Osman Aboul-Nasr
Chemical Technology Center
Fluor
100 Fluor Daniel Drive
Greenville, SC 29680
USA
1 (864) 281-5167
osman.aboul-nasr@fluor.com

Toll Manufacturing: a Perspective from the Tollers View

Mike Ziegler, Arichem LLC, 187 Sloss Industries Road, Ariton, AL 36311

Toll manufacturing is a collaborative effort between a customer and a contract manufacturer. The customer usually only deals with one, maybe two, contract manufactures at any given time. Contract manufacturers may be accommodating several customers at any time, potentially working with double digit numbers, if not more, of customers over the course of a year to avoid equipment and employee idle time.

From the contract manufacturer perspective, every customer brings a unique set of requirements and expectations which may or may not be a good fit in the collaboration. In this presentation, the requirements and expectations of a toll manufacturer will be discussed.

Optimal Operation of Azeotropic Distillation Columns

Salvador Aldrett, Reactive Hazards Assessment, Shell Global Solutions US, 3333 Highway 6 South, Room E-2061, Houston, TX 77082

Heterogeneous azotropes are common in distillation operations involving the removal of a component through the overhead. The azeotrope usually consists of water and other non water miscible organic component such as toluene. A common scheme to separate water from ethanol is to add a third component (benzene) to the column in order to form a low boiling azeotrope between benzene and water and separate the ethanol from the water.

In certain operations, however, it is critical to maintain the solvent to water ratio inside the column in order to prevent falling in the "wrong" side of the azeotrope. This presentation explains how a relatively straight forward separation can go wrong by not watching the solvent to water ratio inside the column. In some instances, running a column "solvent-rich" could cause the solvent to accumulate and bulge inside the column, bulging could have fatal consequences if the accumulating solvent is sensitive in high concentrations.

Study of Treating Ammonium-Nitrogen Wastewater

Qian Wang, Jian Zhou, LuFeng Lv, and Zhanzhu Zhang. RIPP,Sinopec, Beijing, 100083, China

In present research, a novel chemical precipitation technology through struvite formation was developed for removing ammonium-nitrogen from FCC catalyst manufacturing waste water. Comparing with the traditional technology,via this new technology side reactions can be limited, which resulted in higher precipitation efficiency and reagent reaction percent. The pilot-scale experimental results showed that the most important factors influencing the ammonium-nitrogen removal rate are pH value in each steps and the reactants ratio. It showed that for water with 3444mg/L initial ammonia nitrogen when the final pH at 9.0, magnesium ammonium and phosphate molar ratio of 1.15:1:1, the ammonia remnants is under 50mg/L, the removal efficiency is above 99%,and the residual phosphate is very low. Using present technology the complex manufacturing waste water with 1200~1500mg/L initial ammonia nitrogen can be treated in one step accorded with the draining standard. Since the precipitate magnesium ammonium phosphate can be sold as a fertilizer, treating waste water with 1000mg/L ammonia nitrogen will cost 0.36$ per ton via this new technology. So it is a high efficiency and economical method to be put into practice widely.

Redesign of Commercial Process Ion Exchange Purification to Eliminate Hazardous Byproduct Formation

Joseph B. Powell, Ph.D.
Chief Scientist, Chemical Engineering
Shell Global Solutions (US) Inc.
Westhollow Technology Center
Houston, Texas 77082

Introduction

A new commercial process was developed, consistent with the sustainability goals of Shell, to provide reduced environmental footprint in production of a more sustainable product used in manufacture of consumer goods. Following startup, trace production of 1,4-dioxane was observed in a byproduct stream associated with recycle of multivalent cation. While the trace amounts of 1,4-dioxane formed were within plant permits, this molecule is a priority pollutant which is not readily biodegraded, and not readily separable from water using EPA best available technology. According to the principles of sustainability, it was therefore desirable to determine where in the process 1,4-dioxane was being made, and how its formation may be eliminated.

Source Identification

Plant samples indicated the 1,4-dioxane was present only in the acid regenerant stream used to recycle trace multivalent cation leached from the process. This stream was used to regenerate strong, sulfonic acid resin used for multivalent cation removal. Trace organic impurities in the process stream contacting the resin resulted in formation of 1,4-dioxane during regeneration. The phenomenon was subsequently demonstrated in batch laboratory experiments using exaggerated component compositions of the process feed stream. Dioxane formation over strong acid resin has been reported previously.[1]

Process Redesign

Laboratory tests confirmed that trace organic impurities in the process feed stream could ultimately generate a small amount of 1,4-dioxane when contacted with strong acid ion exchange resin under conditions employed across the resin adsorption and regeneration cycle. Weak acid resins (carboxylate) were tested and found to completely mitigate dioxane formation. These resins were however much less effective in multivalent cation removal when tested previously during process design. A program was therefore undertaken to assess under what conditions weak acid resin might be substituted. These resins are known to be effective at multivalent cation removal for pH > 5. Process pH however ranged from 2.5 – 3.5.

Redesign of the ion exchange system used off-line, small-scale techniques that had been previously employed in screening of options for process design. Confidence gained in the combination of these techniques, with mathematical modeling to predict continuous-flow performance, avoided the need to restart large-scale pilot plants in search for a solution. A key parameter in design of adsorption and ion exchange systems is the curvature of the adsorption isotherm.[2] Once the isotherm is defined, analytical models can predict breakthrough performance with sufficient accuracy for use in commercial process design. Small-scale experiments with resin tumbled overnight in vials were used to assess isotherm slope K vs. pH, for the nearly-linear isotherms observed with weak acid resin. Results are shown in Figure 1.

Figure 1: Isotherm slope K as a function of pH for M^{2+} removal on weak acid resin
$K = 0.096 [\exp\{1.81*pH\}]$

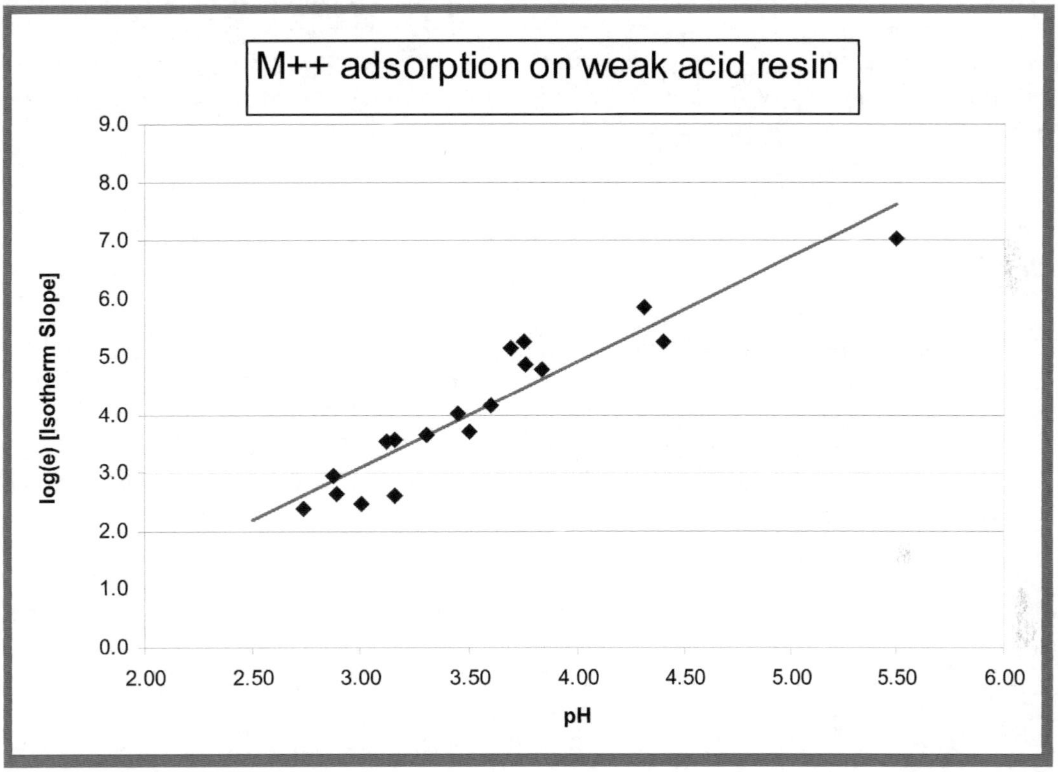

An accurate correlation was required to predict adsorption capacity, given the borderline nature of the design with substitution of weak acid resin.

Retrofit feasibility

Capacity for removal of multivalent cation was only one aspect of overall commercial process requirements. In order to be operable commercially, the resin must provide a minimum mechanical strength, exhibit a fluidization profile which would allow backwash in existing equipment, and provide acceptable pressure drop and mechanical strength in downflow service in existing equipment. It must be capable of being regenerated relative to other contaminants also present, and must not degrade other reactive intermediates in the process. Moreover, multivalent cation removal would no longer be 100% effective, such that the impact of low levels of multivalent cation breakthrough downstream in the process must be assessed. A plethora of small-scale laboratory methods were used for these assessments, including restart of parallel micro-flow reactors to assess downstream impacts. The overall program paralleled that used for initial process development, except for the restart of large continuous pilot plants.

End Result

Weak acid resin was substituted commercially, affording 90% removal of multivalent cation with complete elimination of 1,4-dioxane formation. Substantial process risk was taken in this substitution, as there was no engineering contingency in retrofit design performance with weak acid resin. Correlations developed for redesign are being used for design of pH adjustment systems to effect 100% multivalent cation removal in future operations.

References

1. J. Yamanis and B. D. Patton, Reactions of Ethylene Oxide and Ethanol in the Presence of Amberlyst 15: Products and Product Rate Modeling, *Canadian J. Chemical Engineering, 37*, 297 (197
2. T. Vermeullen, M. D. LeVan, N. K. Hiester, and G. Klein, "Adsorption and Ion Exchange", in *Perry's Chemical Engineer's Handbook*, 6th edition, McGraw-Hill, NY (1984), Section 16

Feed Contamination: We Have Met the Enemy --- It Are Us

Jason V. VanRoeyen, Geismar Manufacturing Facility, Shell Chemical Company, River Road, Geismar, LA 70734, Hilario Garza, HODer, Shell Global Solutions, 3333 Highway 6 South, Houston, TX 77082, and Jonathan H. Worstell, HODER, Shell Global Solutions US, 3333 Highway 6 South, Room E-2061, Houston, TX 77082.

Many chemical processes are comprised of production modules. Coupling two or more such modules may be required to produce a chemical product. Since such modules are connected, the product of one module becomes the feed of a downstream module. Thus, inadvertent contamination of the upstream product leads to manufacturing disruptions in the downstream module. This situation is particularly true when production in the downstream module depends upon a highly reactive; therefore, highly sensitive, catalyst, which generally means an expensive catalyst. There are many analytical methods for quantifying process stream contaminants; however, for highly active catalysts, most such contaminants are lethal at parts per million or less, which is generally at or below the detection limit of the analytical method. This paper discusses the use and limitations of various analytical methods for detecting and quantifying process stream contaminants with regard to catalyst deactivation. For highly active, sensitive catalysts, it is best to use a catalyst performance test to determine the quality of a given process stream. This paper also discusses such a test.

Analyzing Ruptured Pipe Failures

Michael Morimoto, Metallurgical and Materials, Shell Global Solutions, 3333 Highway 6 South, H, TX 77082, Shawna S. Berg, Geismar Manufacturing Facility, Shell Chemical Company, River Road, Geismar, LA 70734, David A. Hampton, Metallurgical and Materials, Shell Global Solutions, 3333 Highway 6 South, Houston, TX 70734, and Jonathan H. Worstell, HODER, Shell Global Solutions US, 3333 Highway 6 South, Room E-2061, Houston, TX 77082.

Ruptured pipes in a chemical processing plant constitute a major safety and environmental hazard. Each such rupture must be analyzed for * why it happened; * how it happened; * can it happen again; * can it happen elsewhere in the plant. Such an investigation involves more than plant personnel. To thoroughly understand a given pipe rupture, metallurgists, process engineers, process chemists, process hazards analysts, and many others must be included in the investigation. This paper discusses a typical pipe rupture with respect to the organization required to investigate such an incident.

AUTHOR INDEX